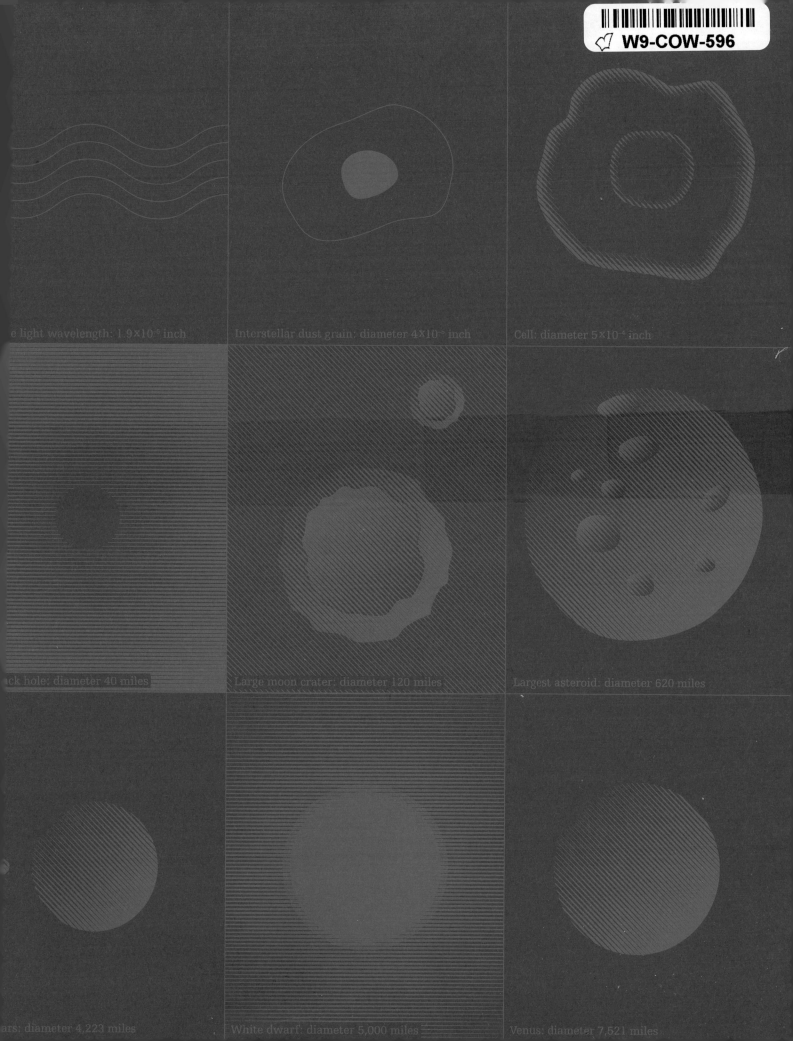

e light wavelength: 1.9×10⁻⁶ inch

Interstellar dust grain: diameter 4×10⁻⁶ inch

Cell: diameter 5×10⁻⁴ inch

ack hole: diameter 40 miles

Large moon crater: diameter 120 miles

Largest asteroid: diameter 620 miles

ars: diameter 4,223 miles

White dwarf: diameter 5,000 miles

Venus: diameter 7,521 miles

THE VISIBLE UNIVERSE

High over the South Pole, collisions between air molecules and charged particles blown from the Sun trigger the shimmering colors of the

aurora australis. Aurorae, which also occur in high northern latitudes, are easily visible to the naked eye.

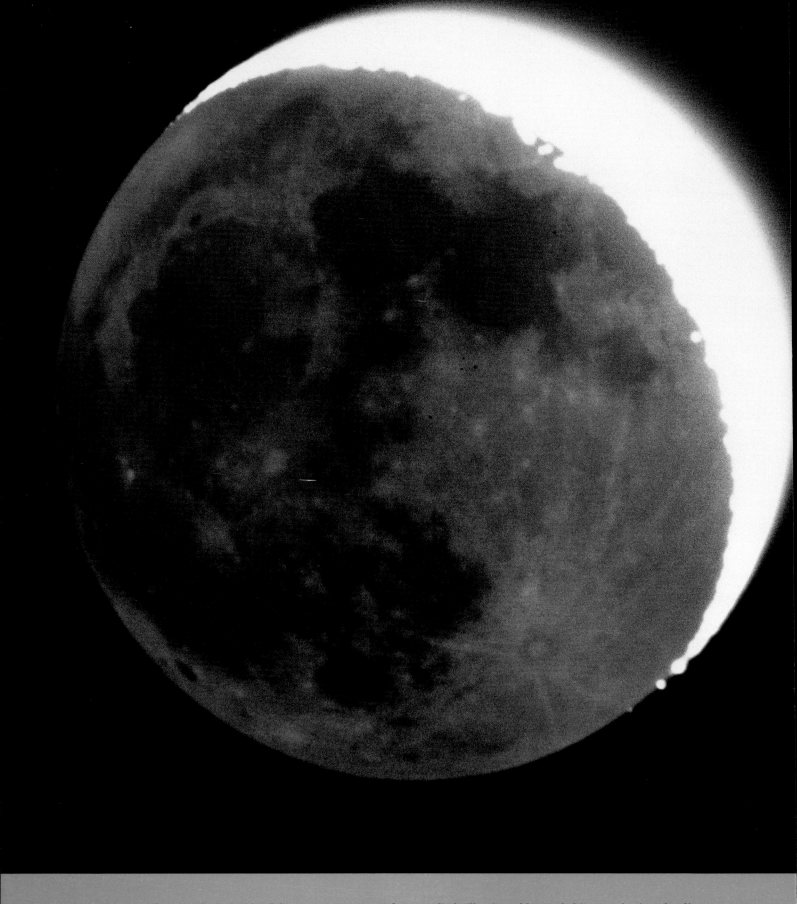

The normally invisible dark side of the Moon, an expanse of craters dimly illuminated by earthshine, nestles in a dazzling crescent.

Comet West trails plumes of gas and dust a quarter of a million miles long in a black-and-white photograph that has been tinted to show the comet's natural blue color.

A color photograph of the Orion nebula, 1,300 light-years away, reveals the magenta glow of hydrogen clouds energized by ultraviolet light

from newborn stars; the naked eye is more sensitive to the nebula's blue-green tones, which show the presence of oxygen.

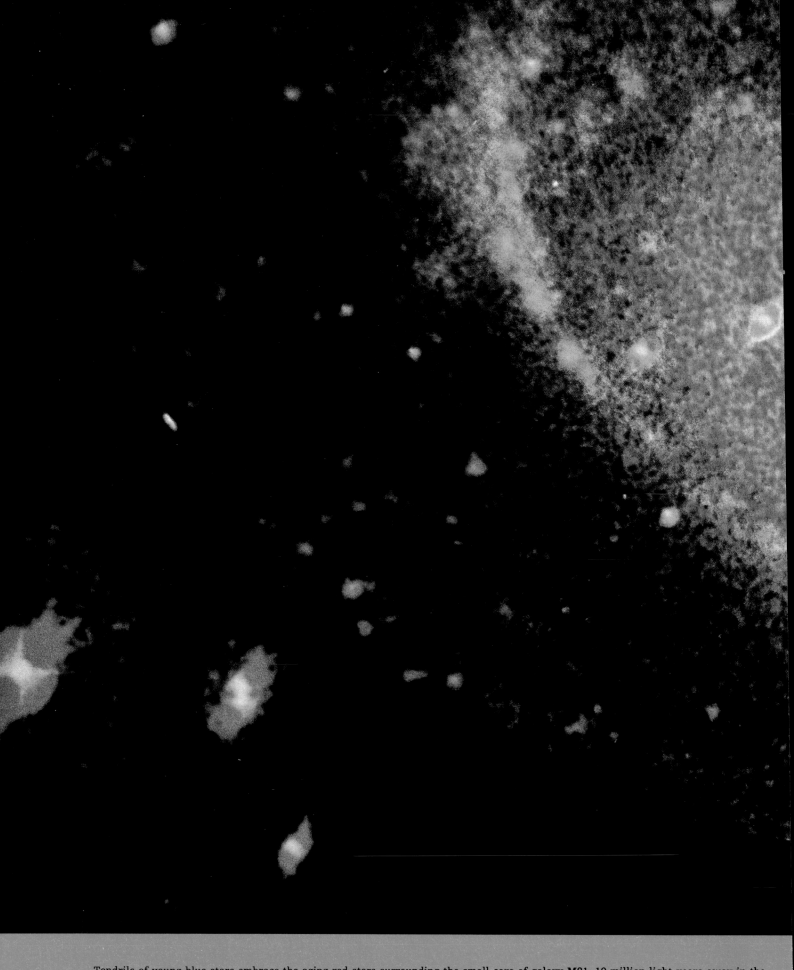

Tendrils of young blue stars embrace the aging red stars surrounding the small core of galaxy M81, 10 million light-years away in the

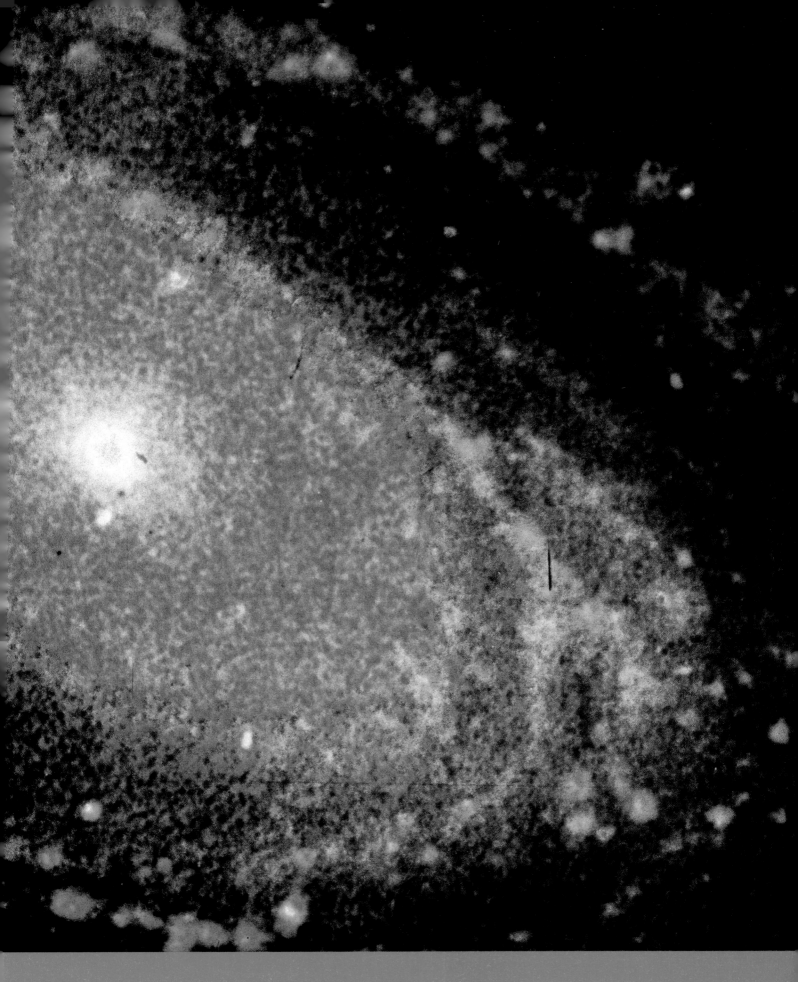

constellation Ursa Major. Colors have been intensified by filtering.

Dominated by two bright elliptical galaxies, this handful of star systems occupies the center of the Virgo cluster, a group of more than a

thousand galaxies 70 million light-years from Earth.

Other Publications:
THE NEW FACE OF WAR
HOW THINGS WORK
WINGS OF WAR
CREATIVE EVERYDAY COOKING
COLLECTOR'S LIBRARY OF THE UNKNOWN
CLASSICS OF WORLD WAR II
TIME-LIFE LIBRARY OF CURIOUS AND UNUSUAL FACTS
AMERICAN COUNTRY
THE THIRD REICH
THE TIME-LIFE GARDENER'S GUIDE
MYSTERIES OF THE UNKNOWN
TIME FRAME
FIX IT YOURSELF
FITNESS, HEALTH & NUTRITION
SUCCESSFUL PARENTING
HEALTHY HOME COOKING
UNDERSTANDING COMPUTERS
LIBRARY OF NATIONS
THE ENCHANTED WORLD
THE KODAK LIBRARY OF CREATIVE PHOTOGRAPHY
GREAT MEALS IN MINUTES
THE CIVIL WAR
PLANET EARTH
COLLECTOR'S LIBRARY OF THE CIVIL WAR
THE EPIC OF FLIGHT
THE GOOD COOK
WORLD WAR II
HOME REPAIR AND IMPROVEMENT
THE OLD WEST

This volume is one of a series that examines the universe in all its aspects, from its beginnings in the Big Bang to the promise of space exploration.

THE VISIBLE UNIVERSE

BY THE EDITORS OF TIME-LIFE BOOKS
ALEXANDRIA, VIRGINIA

CONTENTS

Sheltered by the largest dome of California's Palomar Observatory, the 200-inch Hale Telescope peers out at a skyful of wheeling stars in this multiflash, long-exposure photograph. The Hale mirror, which saw its first light in 1947, continues to make important contributions to investigations of the universe's distant reaches.

ust below the bowl of the Big Dipper in the northern sky, an object glows so faintly that it is invisible to the naked eye. For many years, only the mighty 200-inch Hale telescope atop Palomar Mountain in California could glimpse it—and only when that remarkable instrument was fitted with an electronic camera whose light-gathering ability far exceeds that of the most sensitive film. The object, designated PC 1158 + 4635, is a quasar. Quasars were first detected in the early 1960s by telescopes that collect radio waves rather than light. (The term quasar is a contraction of quasi-stellar radio source.) Soon, however, they revealed themselves at visible wavelengths as well. Hundreds of quasars have been cataloged so far, most of them discovered, like PC 1158 + 4635, by their emanations in the visible portion of the electromagnetic spectrum.

Believed by astronomers to be the centerpieces of young galaxies, quasars are among the brightest objects in the universe. Their perceived faintness is a consequence of their great distance from the Solar System: The nearest one seen to date is about a billion light-years away. PC 1158 + 4635, one of the most remote quasars yet discovered, lies some 14 billion light-years from Earth. Since astronomers believe the universe to be approximately 15 billion years old, the light from this quasar may have started its journey toward Earth less than one billion years after the Big Bang, which marked the beginning of everything. If so, the quasar lies very near the edge of the cosmos— too near the edge, in fact, to jibe with the prevailing theory of the evolving universe, which suggests that galaxies did not begin forming until at least a billion years after the Big Bang. Once again, the act of making itself visible, the universe has coughed up a mystery.

Astronomical history suggests that light will play a role when the answer to this particular puzzle is found. Light has borne to Earth most of what humanity knows about the cosmos: that the planet circles the Sun; that the Solar System is rich in other planets, their attendant moons, and other rocky phenomena; and that the universe is built of huge collections of stars called galaxies, all rushing headlong toward an uncertain end; as well as other facts too numerous to recount.

Since the seventeenth century, the optical harvest of information about the universe has been gathered in by telescopes. But prior to the invention of that revolutionary tool, astronomers solved many a cosmic conundrum with a most imperfect instrument—the human eye. Shaped by millions of years of evolution to view nearby objects illuminated by the Sun, the eye is rather poorly suited to detecting starlight diminished after a long journey to Earth. Nonetheless, it can see much in the skies to ponder. The Sun and Moon, of course, cannot be overlooked. On a cloudless night, far from the glare of cities, some 6,000 stars appear strewn across the heavens. One of the brightest of these pinpricks of light, often appearing near the horizon at dusk, is not a star at all but the planet Venus reflecting light from the Sun. At one period or another during the year, Mars, Saturn, Jupiter, and Mercury are all similarly visible without a telescope. And comets trailing long plumes of material show themselves from time to time.

EARLIEST OBSERVATIONS

Careful observation of the heavens suggested to ancient astronomers that stars moved as if painted on a revolving dome encompassing the Earth. More than 4,000 years before the birth of Christ, Sumerian astronomers, living in the area known today as Iraq, tracked the Sun, the Moon, and the planets across the sky. To do so, they established the practice of dividing a circle into degrees, minutes, and seconds, measures that have been used ever since to specify the positions of stars and other objects in the firmament. Egyptian astronomers later used the stars to keep track of time. Having observed that the Pole Star in the north stands still in the sky, they hung a plumb bob so that the string appeared to intersect the star. Then they marked the passage of the night as other stars on the revolving dome overhead crept past the taut string. Aligning two plumb bobs with the Pole Star allowed a useful position-reckoning line, now called a local meridian, to be drawn on the floor of an observatory.

The Greeks used light from the Sun to take the measure of Earth. By applying geometry and trigonometry to the analysis of the varying lengths of shadows cast by the Sun at different points on Earth's surface, they deduced the planet's spherical shape and calculated its diameter. Eclipses were also used for measurement. A solar eclipse in the second century BC appeared total when viewed from a vantage point in Alexandria, Egypt; but about 750 miles north in Byzantium, at the mouth of the Black Sea, the same event obscured only 80 percent of the Sun. From these observations, the Greek astronomer Hipparchus attempted—not very successfully—to compute the distance between Earth and the Sun. His result, just under two million miles, was about two percent of the true value.

Hellenic astronomy culminated with the Alexandrian scholar Ptolemy. Among his many contributions was the invention of an instrument called a quadrant—an arc amounting to one-fourth the circumference of a circle. At what would be the center of the circle was a pivoting rod or pipe that served as a pointer. By sighting along the pointer aimed at a star and noting its position on a scale engraved on the arc, the altitude of the star above the horizon could be found. When paired with the angle between the lo-

PRECISION TOOLS FOR MEASURING THE HEAVENS

Long before telescopes began providing clearer views of the heavens, astronomers were taking the measure of celestial objects with such devices as the quadrant *(middle)*, which pinpointed the altitude of an object above the horizon. Using the naked eye and instruments like these, Tycho Brahe made observations so precise that they led Johannes Kepler to formulate his laws of planetary motion and to conclude that all the planets revolve around the Sun.

With the advent of telescopes, astronomers could use quadrants and newer tools such as micrometers and photometers to determine the position, dimensions, and brightness of heavenly bodies with even greater accuracy. The micrometer *(near right)*, conceived around 1639 by English astronomer William Gascoigne, served to measure the angular diameter of relatively nearby bodies such as the Sun, Moon, or planets. Starting in the mid-1700s, the photometer *(far right)* was used to gauge an object's brightness by comparing it to something whose brightness was known, such as a candle or a very bright (and therefore long-studied) star such as Polaris.

Micrometer. Mounted between the objective and the eyepiece of a telescope, a so-called filar, or fine hair, micrometer consisted of a fixed wire *(1)*, a movable wire *(2)*, and a cross hair *(3)* for orientation. An object's angular diameter was measured by lining one edge of it against the fixed wire and turning a dial to shift the moveable wire rightward until it touched the other edge. An astronomer determined angular measurement from the number of turns, or parts of turns, the dial made.

cal meridian and the star, the altitude fixed the star's location in the sky.

Seeking to express the order that he believed to be implicit in his own celestial observations and those of others—all taken by eye with crude apparatus—Ptolemy proposed a model for planetary motion that would prevail for 1,500 years. In essence, he envisioned the universe as a system of perfectly transparent, concentric, and steadily rotating spheres. The Earth stood at the center; stars occupied the outermost sphere. Between lay a number of additional spheres, one for the Moon, another for the Sun, and one for each of the planets. Other unoccupied and interlocking spheres, turning at different rates, meshed with the planetary spheres like a huge set of gears. These auxiliary spheres were necessary in order to account for the erratic motions of some celestial bodies as they appeared to ancient astronomers. The planets, for example, traveled at changing speeds, belying the supposed steady rotation of their spheres. Sometimes they even seemed to reverse course, a phenomenon known as retrograde motion. Ptolemy's system, though un-

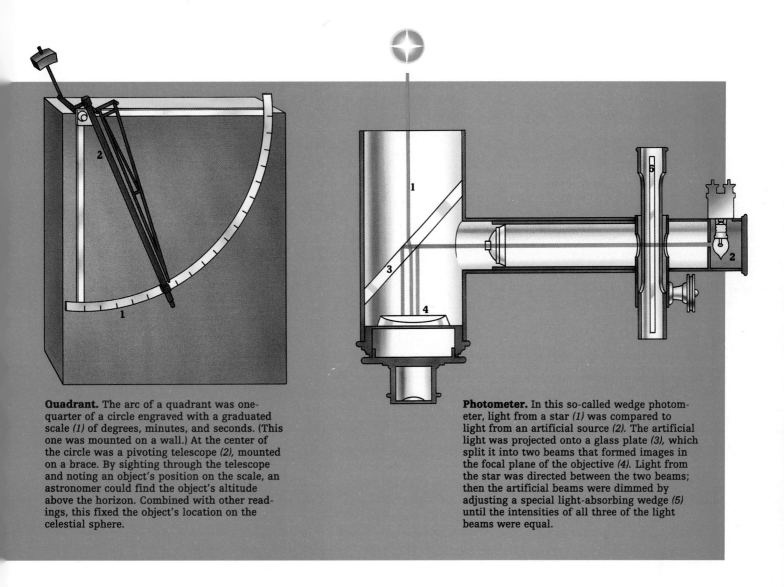

Quadrant. The arc of a quadrant was one-quarter of a circle engraved with a graduated scale *(1)* of degrees, minutes, and seconds. (This one was mounted on a wall.) At the center of the circle was a pivoting telescope *(2)*, mounted on a brace. By sighting through the telescope and noting an object's position on the scale, an astronomer could find the object's altitude above the horizon. Combined with other readings, this fixed the object's location on the celestial sphere.

Photometer. In this so-called wedge photometer, light from a star *(1)* was compared to light from an artificial source *(2)*. The artificial light was projected onto a glass plate *(3)*, which split it into two beams that formed images in the focal plane of the objective *(4)*. Light from the star was directed between the two beams; then the artificial beams were dimmed by adjusting a special light-absorbing wedge *(5)* until the intensities of all three of the light beams were equal.

wieldy, enabled a competent mathematician to predict future positions of the planets with acceptable accuracy.

The beginning of the end for this overly complex framework came in May 1543, when the Polish churchman Nicolaus Copernicus published his ideas for a celestial model that placed the planets, including Earth, in motion around the Sun. This was heresy; the Catholic Church had adopted the Greek system, pagan though it was, as Christian doctrine. Perhaps fortunately, Copernicus died shortly after his Sun-at-the-center hypothesis was published; he did not have to defend it. He probably could not have persuaded anyone that he was nearer the truth than Ptolemy, since in Copernicus's time, the measurements of the heavens that might have proved him correct were themselves none too trustworthy. The source of the difficulty was the imprecision of the devices used to establish the positions of celestial bodies. For example, Copernicus's quadrant was little better than Ptolemy's, which had an accuracy of about 10 arc minutes—one-sixth of a degree.

Some twenty years later, the deficiencies of the quadrant and other astronomical tools became all too apparent to a Danish practitioner of the science named Tycho Brahe. Observing Jupiter and Saturn with a sextant, an instrument invented at the turn of the sixteenth century to measure the angle between a pair of heavenly bodies, he found that the time of their closest approach in the sky differed by several days from that calculated with the most precise celestial-motion tables then available. Attributing the discrepancy to faulty observations, he resolved to make the best measurements of the planets ever achieved.

IMPATIENT TYCHO

Tycho Brahe was born of the Danish nobility. Taken from his parents and raised by an imperious uncle who simply wanted a child, he grew up to be something of a black sheep. At age twenty, for example, he dueled with a fellow student at Rostock University over a fine point of mathematics, losing part of his nose in the fight. (Tycho fashioned a substitute for the severed portion from copper, silver, and gold, painting it the color of flesh and fixing it to his nasal stump with adhesive.) He also rebelled at studying rhetoric and philosophy as preparation for becoming a statesman, a traditional family occupation. Instead, impressed by astronomers' ability to forecast a solar eclipse, he chose to join the ranks of stargazers.

Tycho intended to improve the accuracy of astronomical measurements by building better tools, but he was thwarted by lack of money until 1572. That year, the twenty-six-year-old astronomer gained wide recognition after noting a brilliant light in the night sky. Thinking that he had discovered a new star, Tycho wrote a pamphlet about it the next year entitled "De Nova Stella." Although he did not know it at the time, Tycho had observed the explosion of a star, an event now known as a supernova. Largely as a consequence of his booklet, which overthrew contemporary notions that stars were unchanging and fixed in number, he came to the attention of the Danish king, Frederick II. In 1576, the king granted him the 2,000-acre island of Hveen, near Copenhagen, and supplied the funds necessary to build an observatory. The result was the greatest such facility of its time.

Tycho equipped his observatory with better astronomical instruments than the world had seen. "Better," during the late sixteenth century, meant bigger—outsize sextants, for example, and a celestial globe five feet in diameter that showed positions of the stars. Perhaps most impressive among Tycho's tools was a huge quadrant. The greater the circumference of the circle from which a quadrant was constructed, the more precisely the scale could be subdivided into minutes and seconds. Tycho's instrument, with a radius of seven feet, was so large that he mounted it on a wall for stability. With this mural quadrant, as it was known, Tycho could take measurements as small as one-half minute of arc, up to twenty times better than earlier instruments.

Such precision had the potential to upset age-old but erroneous theories about the heavens. Since antiquity, for example, astronomers had believed that comets were atmospheric phenomena, like rainbows. Tycho's observations contradicted this idea, irrefutably placing comets well beyond the Moon. The Danish astronomer also cataloged the positions of 777 stars and, over a period of two decades, tracked the movement of Mars and other planets with unprecedented accuracy.

Along with possession of Hveen had come responsibility for the welfare of its tenants—a responsibility that Tycho disdainfully ignored. King Frederick tolerated this behavior, but his successor, Christian IV, was less lenient. After a dispute over the issue with the new monarch, the astronomer, having become somewhat bored with his work anyway, left Denmark in 1597. Taking his tabulations of planetary data with him, he settled in Prague, there to become Imperial Mathematician to Holy Roman Emperor Rudolph II. He died suddenly in 1601, after excessive consumption of beer and food at a banquet of Czech nobles.

Tycho went to his grave unaware that hidden within his careful observations of the planets lay data to support a theory of planetary motion more Copernican than Ptolemaic—and more correct than either. That discovery fell to Tycho's assistant in Prague, a brilliant young German astronomer named Johannes Kepler, who assumed the post of Imperial Mathematician after Tycho's passing. Kepler's analysis of his mentor's logbooks revealed that Mars and the other planets, including Earth, revolve in ellipses around the Sun and that the Moon traces an ellipse around Earth. Tycho's legacy also enabled Kepler to formulate laws of planetary motion that explain not only the changing speeds of planets but also their occasional retrograde motion, a consequence of a planet's motion relative to Earth's.

Kepler's proposals, though insightful refinements of the Copernican model,

TELESCOPIC EVOLUTION

From Galileo onward, astronomers have tinkered with the telescope's components, striving to see ever deeper into the universe. During the first 340 years of this quest, highlighted on the following pages, lensed telescopes, or refractors, vied for primacy with mirrored reflectors as scientists tried to solve the problems associated with gathering and focusing the faint light of the stars. For example, the earliest lenses magnified distant objects severalfold but yielded blurry images. Lengthening the focal length of the objective lens sharpened the image but created a new problem: increasingly unwieldy support structures. By the end of this period, advances in design and technology had brought mirrors to the fore, culminating in the 200-inch Hale.

could not be proven by the data at the astronomer's disposal. As accurate as the information was, it still allowed for alternate, albeit tortured, interpretations. To make measurements of definitive precision, astronomers needed to improve their eyesight. Fortunately, in 1609, the same year that Kepler published two of his laws of planetary motion (he would later formulate a third), forty-five-year-old Galileo Galilei of Padua was turning his first telescope toward the heavens.

AN OPTICAL REVOLUTION

Galileo and similar-minded investigators of his era benefited from centuries of progress in practical optics. In Egypt, for example, glass had been made from sand since 3500 BC. The earliest recorded studies of glass's ability to focus light by bending or refracting it came more than thirty centuries later, from the Greek geometer Euclid. A medieval Arabic physicist named Alhazen was the first to put forward a theory of refraction. After his death, thought to have occurred in 1038, his work on the subject was translated into Latin, the language of science in Europe. By the late 1200s, the first spectacles maker had opened shop in Italy. Soon, opticians were common throughout Europe, and shortly after the turn of the seventeenth century, Hans Lippershey, a producer of eyeglasses in the Dutch city of Middleburg, was assembling primitive spyglasses that used the refractive properties of glass to magnify distant objects.

As Galileo later recounted, "a report reached my ears that a Dutchman had constructed a telescope" consisting of a tube containing two lenses. Galileo's whole life had prepared him to improve and exploit the invention. A mathematics professor and experimental physicist, he had already advanced the

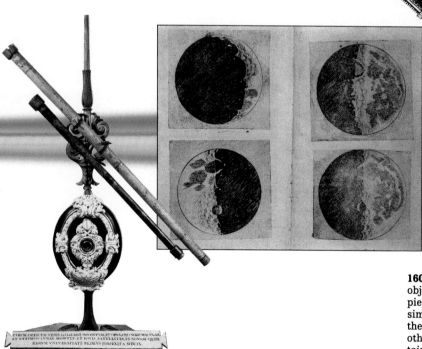

1609 Galileo Galilei paired convex objective lenses and concave eyepieces in narrow tubes to make simple telescopes *(far left)*. He used the instruments to discover, among other things, craters and mountains on the Moon *(above, left)*, Jupiter's four largest satellites, and mysterious objects at Saturn, which he called "appendages."

science of dynamics—the study of motion—with experiments disproving the notion that heavy bodies fall faster than light ones. He was also interested in astronomy and the Copernican theory. As a skilled designer of mechanical equipment and delicate apparatus, he had by 1609 patented an improved water mill for agricultural irrigation, built a pulsilogium (an instrument to time a patient's heartbeat), and engineered a geometrical-and-military compass (a calculating device and forerunner of the slide rule that also was used to measure angles in the landscape of a battlefield). Finally, to supplement his income from teaching, Galileo made instruments for sale. Seeing a financial opportunity in the new telescope, which had obvious military applications, he built one.

Using materials at hand, he mounted a convex lens and a concave lens at opposite ends of a lead tube that held the two disks of glass in alignment and excluded stray light. Then, "bringing my eye to the concave lens," he wrote, "I saw objects satisfactorily large and near." The convex lens in his telescope came to be called the objective; it formed an image of a distant object for viewing through the concave eyepiece, which magnified the scene.

Pleased with his work, Galileo a week later assembled a second telescope, in which the lead pipe was replaced by a tube rolled from tinned iron plate, probably soldered along its length, and covered in crimson cloth. He then hastened to Venice to show his newest creation to Venetian merchants and military men. In a letter to his brother-in-law in August 1609, Galileo wrote:

1611 By projecting the Sun's image onto a sheet of paper mounted behind the eyepiece of his refracting telescope *(below, left),* the Jesuit astronomer Christoph Scheiner of Germany could safely study sunspots *(below),* which he assumed to be small planets that orbited the Sun.

1647 To reduce color blurriness, or chromatic aberration, Poland's Johannes Helvelius tried shallowly curved objective lenses and long focal lengths, using instruments 8 to 12 feet long to produce detailed maps of the lunar landscape *(below).* He mounted even longer telescopes on his roof *(left).*

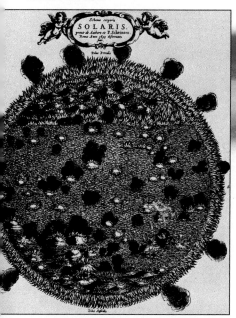

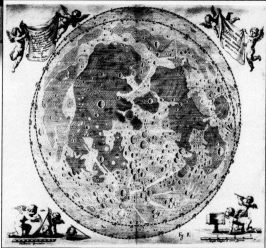

"There were a great many gentlemen and senators who, although old, climbed the stairs of the highest belfries of Venice several times to look for sails on the sea, and ships so far away that, even under full sail, two hours and more went by before they could be seen" without the aid of his telescope.

A CONCERTED EFFORT

In Padua, Galileo already employed a coppersmith to build instruments under his direction. Now he added glassworking facilities that had equipment for grinding and polishing lenses to the desired shape, or figure. The first difficulty in lens making was to obtain suitable blanks, the disks of glass from which lenses are ground. Toward this end, Galileo befriended Paduan master glassworker Girolamo Magagnati, instructing him to send him the best glass that he could make. He dealt with others, such as Maestro Antonio of Murano, an island near Venice, through intermediaries. Most of the blanks they offered were too flawed with streaks and bubbles for Galileo's purposes. He also sought finished lenses and placed orders with artisans throughout Italy, often with disappointing results. For example, among a batch of 300 lenses submitted by the much-respected Girolamo Bacci of Murano, only three were suitable for a telescope, and even they were imperfect.

Nevertheless, this careful attention to materials and workmanship proved worth the effort. As early as September 1609, a Tuscan visiting Venice reported home that Galileo's telescopes surpassed others that were offered for

1682 To avoid the buffeting of wind on ever-longer wooden telescope tubes, the Dutch astronomer Christiaan Huygens began building "aerial" refractors in which he held the lenses in place with string *(below)*. He used one such instrument to make a drawing of Saturn *(below, right)*, whose ring plane he had identified in 1656.

1660s Sir Isaac Newton discovered that light passing through a prism *(above)* is split into its constituent colors because of the unequal bending of different wavelengths. Concluding (wrongly, it turned out) that the problem of chromatic aberration was unavoidable in lensed telescopes, he built the first successful reflector *(right)* with a 1.3-inch primary mirror.

sale, mostly because of the "good quality of the material of the lenses."

Besides making heroic efforts to obtain suitable glass for his telescopes, Galileo steadily improved the design. Like other opticians, he used lenses that were spherical in curvature because they were the simplest to shape. He soon came to favor subtly rounded lenses over those having a more pronounced curve. The flatter lenses had longer focal lengths, focusing an image farther behind the glass. They therefore required longer tubes than the others, but they produced clearer images. For additional clarity, he learned to make objective lenses larger in diameter than he intended to use. "I can grind them more accurately," he wrote, "since on a more spacious surface it is possible to maintain the shape much better than on a small one." He then masked the outer portions of such lenses. In one telescope with an objective two inches in diameter, he covered all but the central inch or so, thus sacrificing some of the lens's potential light-gathering power for additional image clarity.

Galileo improved his telescopes mechanically as well as optically. For example, he devised a stand to keep the instrument steady in order to "avoid the wavering of the hand." He also designed a telescope made from a pair of concentric tubes. One tube contained the eyepiece and slid into the other, which held the objective. This arrangement permitted fine adjustment of the distance between the lenses to achieve the best image possible.

1721 English astronomer John Hadley's compact reflector *(below)* had a 6-inch parabolic mirror and a 5-foot focal length. With this instrument, he produced drawings—such as the one below, right, depicting the dark gap in Saturn's ring—as sharp as those made using a more unwieldy refractor with an 8-inch lens and a 120-foot focal length.

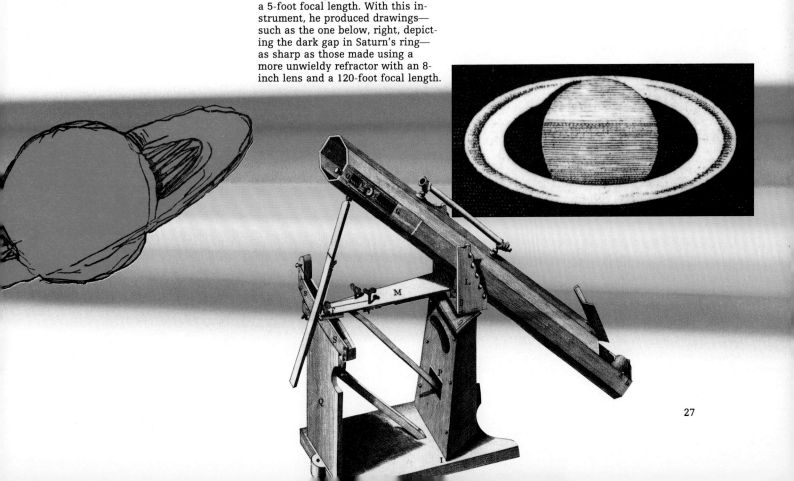

27

Though potential customers for these telescopes were attracted chiefly by their mercantile and military potential, Galileo valued the instruments for their promise in the field of astronomy. Within months of building his first telescope, he began studying the night sky.

Fascinating discoveries were there for the gathering, like rare seashells strewn across an untrodden beach. Between 1610 and 1612, Galileo observed craters and mountains on the Moon, proving that it is a world bearing kinship to Earth. He saw that sunspots lie on the surface of the Sun and are not objects passing in front of it, thus disproving the teaching, handed down from the Greeks, that the Sun is a perfectly unblemished sphere. He determined that the Sun rotates. He observed that the disk of Venus undergoes phases like those of the Moon. He noticed four satellites revolving around Jupiter and named them the Medicean stars, after his patron the grand duke Cosmo di Medici. Galileo also detected something near Saturn that he called appendages of the planet. And, while making a list of the "stars" in Jupiter's vicinity, he sighted the planet that more than two centuries later would be named Neptune.

The cumulative weight of these and other discoveries, made possible only by the telescope, fixed Galileo's conviction that Earth circles the Sun. Holding such beliefs was risky in staunchly Catholic Italy. For espousing this heresy,

1782 To avoid the dimming effect of a secondary mirror, William Herschel built a reflecting telescope *(bottom)* with a 12-inch mirror that focused light directly into an eyepiece near the top of the 20-foot-long tube. From this perch, Herschel did much of his "star gauging"—counting the stars in various directions to build a map of the Milky Way *(below).*

1758 Father and son opticians John and Peter Dollond (whose ornate business card is shown below) patented doublet lenses—concave and convex lenses joined together—with which they built telescopes *(left)* free of chromatic aberration.

28

Galileo was brought before the Inquisition in Rome in 1633, where he was made to recant and to deny that Earth moves. For the next nine years, until his death in 1642, he lived under house arrest, continuing his scientific investigations to the end.

EXCELLENT INSTRUMENTS FROM ITALY

Galileo was by no means the only instrument maker to experiment with Lippershey's invention, but his telescopes set the standards in his day. The instruments produced erect (right side up), bright images. His first refractor magnified objects three times; others he built made things appear as much as thirty times larger than they looked to the unaided eye. Even so, the telescopes had major shortcomings. To begin with, the fields of view were very small. Two of his telescopes, for example, were able to observe a region only one-quarter of a degree in diameter, half the width of the full Moon. Images were blurry and often marred by colored halos.

Ideas for improvements came from the versatile Johannes Kepler. Within months of Galileo's earliest successes with telescopes, Kepler learned of his skill in making these instruments. Writing to him in early 1610, Kepler re-

1826 The German lens maker Joseph von Fraunhofer built a 9.5-inch refractor *(bottom)* at the Dorpat Observatory in Estonia for astronomer F. G. W. von Struve, who used it to resolve 2,200 binary stars (1,100 pairs). Aided by a clock-driven equatorial mount that allowed the telescope to track stars across the sky, Struve made a celestial map *(below)* showing the swath of the Milky Way.

1845 With the "Leviathan of Parsonstown," a 72-inch metal mirror that weighed 4.4 tons *(bottom)*, Irish astronomer William Parsons, the earl of Rosse, resolved the spiral form of some nebulae—now known to be galaxies outside the Milky Way—and made detailed drawings such as this one of the Whirlpool galaxy M51 *(right)*.

marked that the refractors available in Prague were very poor: "The only one which I have gives a twentyfold enlargement, but the light is very weak."

The same year, Kepler wrote *Dioptrice*, a volume of 141 propositions, rules, and examples that constituted the first mathematical analysis of lenses and refraction. He had studied the Arabic scientist Alhazen's works on optics in the hope of correcting an advancing nearsightedness—as well as the multiple vision that had plagued him since childhood. In his book, he discussed the optical properties of the eye, proposing a new kind of refractor to eliminate some disadvantages of the Galilean telescope. By making use of a convex eyepiece, for example, Kepler's design offered a larger field of view than Galileo's instruments.

Kepler never implemented his ideas, but others did, and by 1617 the Jesuit astronomer Christoph Scheiner had employed a telescope with a convex eyepiece to study sunspots. In the 1640s, the Neapolitan astronomer Francesco Fontana used a Keplerian refractor to examine the Moon and planets.

1852 Using the 15-inch doublet refractor at the Harvard College Observatory *(below)*, John A. Whipple made one of the earliest celestial photographs, a daguerreotype of the Moon *(bottom)*.

1897 Built by Alvan Clark and Sons, the 40-inch refractor at the University of Chicago's Yerkes Observatory *(below)* is the world's largest. Key to studies of stellar motions requiring at least twenty years' observation, the 40-inch gave early evidence that the Milky Way, Earth's star-studded home *(right)*, is a spiral galaxy.

By the middle of the seventeenth century, such telescopes had supplanted Galilean refractors. Yet two weaknesses remained—an overall blurriness and colorful halos around bright objects. Kepler had found that blurred images result from a lens's spherical shape. Light from an object on the optical axis of the lens—a line perpendicular to its center—passing through the edge of the lens is focused at a point slightly closer to the glass than light coming through the middle. Images of objects elsewhere in the lens's field of view suffer additional degradation because light from them passes asymmetrically through the lens. Kepler had recognized that this defect, known as spherical aberration, could be all but eliminated with an objective shaped like the lens of the eye, which approximates the shape of a curve called a hyperboloid instead of a sphere. By focusing more light to the same point, a hyperboloid objective makes brighter images of objects near the optical axis, even though it actually diminishes image quality of targets near the edge of the lens. (The villain is an aberration called a coma, which, as it happens, tends to be less

1918 Completed during World War I, Mount Wilson's 100-inch Hooker reflector later enabled Edwin Hubble *(below)* to discern a type of star called a Cepheid variable in the great nebula in Andromeda *(bottom)*. The discovery let him make unprecedented distance measurements, proving that spiral nebulae, once thought part of the Milky Way, are in fact galaxies in their own right.

1947 With four times the light-gathering power of the Hooker 100-inch, the 200-inch Hale mirror at Palomar Observatory *(below)* yields detailed photographs like this one of the Whirlpool galaxy *(below, right)*. Within a few years of its installation, the telescope helped in recalibrating Edwin Hubble's distance scale by proving that Andromeda lies 2.2 million light-years away—more than two times farther than previously believed.

detrimental in spherical lenses than in hyperbolic ones.) Such lenses were costly, however, because they could not be readily fabricated with seventeenth-century lens-making equipment. Even today, grinding hyperbolic lenses is rarely undertaken. As for the halos, the result of a defect called chromatic aberration *(page 36)*, a remedy would not be found for more than a century.

During that interval, astronomers achieved sharper images and minimized halos as best they could by following Galileo's lead and using flatter objective lenses. The benefits—namely, reduced spherical and chromatic aberration—were paid for with longer telescope tubes. In the 1650s, Christiaan Huygens, a Dutch astronomer, used telescopes up to 123 feet in length to investigate the "appendages" that Galileo had glimpsed next to Saturn. Huygens could see with his longest apparatus that they actually formed a ring around the planet.

Telescopes of such dimensions were far from satisfactory, if for no other reason than that they were difficult to point. Often built without an enclosure to offer protection from the elements, they were susceptible to the slightest zephyr. An Englishman who looked through one of these telescopes in 1720 found it dismayingly shaky. He concluded that "not many good observations can be made with a glass of 123 feet in the open air."

CLUES TO A SOLUTION

As astronomers struggled with these marginal instruments, the principles for making more compact telescopes lay right under their noses. One, known since Galileo's time, was that the refractive power of glass—the degree to which it bends light—depends on the material's composition. The other, put forward in 1666 by Isaac Newton, was the explanation of how a prism turns white light into a rainbow of color. Still in his twenties—and yet to father the universal law of gravitation, the basic laws of motion, and the branch of mathematics known as calculus—Newton showed that wedge-shaped blocks of glass called prisms did not infuse sunlight passing through them with color, as most scientists thought, but merely spread apart colors that were already present by bending certain light rays more than others. Lenses, being a form of prism, had a similar effect on starlight as it passed through a telescope. And thus was found the source of the telescope's rainbow-hued halos.

In 1733, English lawyer and amateur scientist Chester Hall had the bright idea of combining these principles to correct the chromatic faults of a convex spherical telescope objective with the compensating faults of a concave lens, creating a compound called a doublet. He also thought to craft the convex part of the doublet from so-called crown glass, or common window glass, which is not very refractive, and the concave element from flint glass, a highly refractive type made by adding lead oxide to the recipe for crown glass. When he positioned the concave lens close behind the convex one, the result was a sharp, nearly halo-free image.

The pursuit of optical perfection has been hampered by uncertainty about the makeup of light itself. In the late seventeenth century, for example, Isaac Newton conducted a series of experiments suggesting that light consists of infinitesimal—and undetectable— particles. Dutch astronomer Christiaan Huygens dissented, proposing that light travels as waves rather than particles. By 1752, Benjamin Franklin's assessment of his own inquiries into the matter seemed a fair summary of progress in understanding the phenomenon: "I must own I am much in the dark about light."

Further study by British physicists Thomas Young and James Clerk Maxwell helped make the wave theory of light dominant during the 1800s. With the birth of quantum physics in the twentieth century, however, the particle theory was revived. Astronomers today remain leery about defining the composition of light, but on its behavior they do agree: Light comports itself both as a rapidly oscillating electromagnetic wave and as a stream of particles, called photons, that often interact with the atoms of a substance. Both a wave and a photon of starlight can be thought of as tracing a line—a light ray—from Earth back to their celestial origin.

Awareness of light's dual nature has led astronomers to a clearer understanding of how it is generated in a star (box, right) and how it can best be collected and enhanced by telescopes on Earth. Light travels through space, they believe, like a three-dimensional version of the familiar ripples that are created when a stone falls into a pond. By the time this expanding sphere reaches Earth, it has spread out so much that the portion striking the planet forms only a tiny, nearly flat patch, called a wavefront.

Although the unaided eye can detect the wavefronts than those visible in the sky to materialize in the instrument's eyepiece. An instrument of that aperture can also resolve details ten times smaller than those perceptible by humans; someone whose eyes enjoyed such resolving power could easily read the text on this page from a distance of forty feet.

Astronomers and physicists gauge light by its wavelength (the distance between two successive wave crests) and frequency (the number of crests that pass a given point in one second). Because frequency is inversely proportional to wavelength, light of the highest frequencies has the shortest wavelengths.

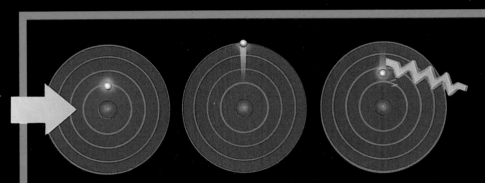

The sequence above traces the genesis of a photon, or packet of light energy, at the heart of a distant star. Orbiting the nucleus of an atom in the star, an electron (left) gets a jolt of energy (yellow arrow) whenever it receives radiation or collides with a neighboring atom. This influx of energy boosts the electron to a higher orbit (center), but the electron's natural tendency to seek its lowest energy state causes it to drop back down (right). As the electron falls to this lower level, it releases its excess energy as a photon of light (green zigzag). An electron dropping between two given orbits will always produce a photon of one specific wavelength.

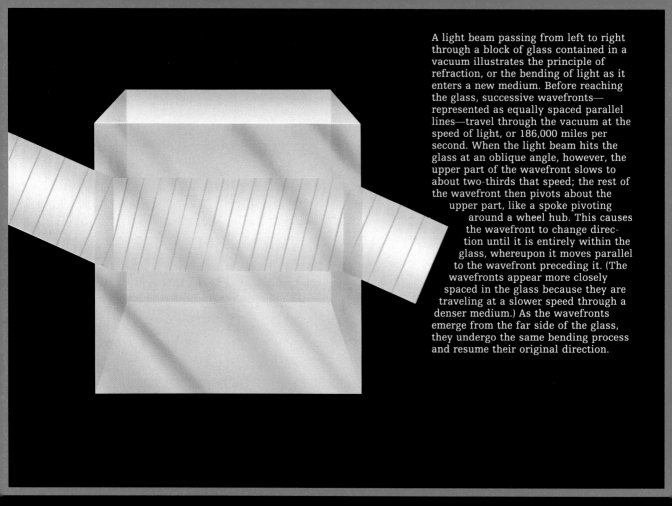

A light beam passing from left to right through a block of glass contained in a vacuum illustrates the principle of refraction, or the bending of light as it enters a new medium. Before reaching the glass, successive wavefronts—represented as equally spaced parallel lines—travel through the vacuum at the speed of light, or 186,000 miles per second. When the light beam hits the glass at an oblique angle, however, the upper part of the wavefront slows to about two-thirds that speed; the rest of the wavefront then pivots about the upper part, like a spoke pivoting around a wheel hub. This causes the wavefront to change direction until it is entirely within the glass, whereupon it moves parallel to the wavefront preceding it. (The wavefronts appear more closely spaced in the glass because they are traveling at a slower speed through a denser medium.) As the wavefronts emerge from the far side of the glass, they undergo the same bending process and resume their original direction.

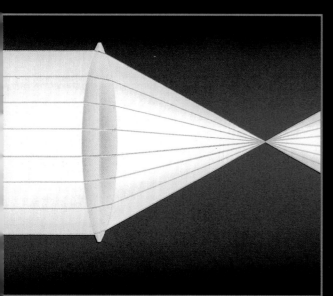

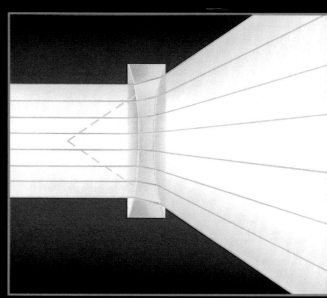

The thin edges of a lens that is convex on both its front and back sides bend incoming parallel rays of light at a greater angle than does the lens's thick center; as a result, the light rays converge

A concave lens, with a thin center and thick edges, causes parallel light rays to diverge. To an observer, the rays appear to have originated from a so-called virtual focus *(dotted lines)* on

THE ART OF BENDING LIGHT

Although light zips unimpeded through the vacuum of space at 186,000 miles per second, it traverses other, denser media at a comparatively sluggish pace: In the glass that makes up most optical telescope lenses, for example, light moves at a speed of approximately 124,000 miles per second.

How the light behaves as it passes from one medium to another depends on the angle of the encounter, called the angle of incidence. When light hits the surface of glass at right angles, it travels straight through the medium and continues on its course. But when the incident angle is oblique *(opposite, top)*, the light bends toward the normal—that is, a line perpendicular to the surface—as it passes through.

Since the early seventeenth century, astronomers and lens makers have harnessed this bending of light—called refraction—to produce optical telescopes known as refractors *(below)*. Such instruments incorporate two basic lens types: Convex lenses use refraction to converge parallel rays of light to a focus; concave lenses employ the same effect to make parallel rays diverge. The distance between a lens and its focal point is called the focal length, and the magnifying power of any refracting telescope is the ratio of the focal length of the light-collecting (or objective) lens to the focal length of the eyepiece lens.

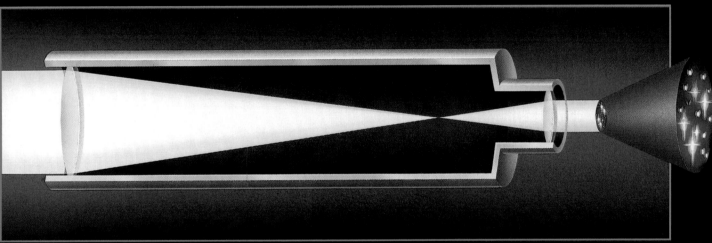

A refracting telescope of Keplerian design *(above)* relies exclusively on convex lenses. The larger lens, called the objective, brings light to a focus inside the telescope tube, while the smaller lens, called the eyepiece, magnifies the image so produced.

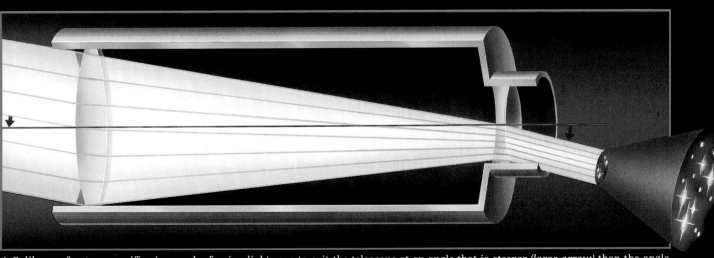

A Galilean refractor magnifies images by forcing light rays to exit the telescope at an angle that is steeper *(large arrow)* than the angle at which they entered *(small arrow)*. The convex objective lens brings the light nearly to a focus, where it is refracted, or bent, by the concave eyepiece to form parallel rays.

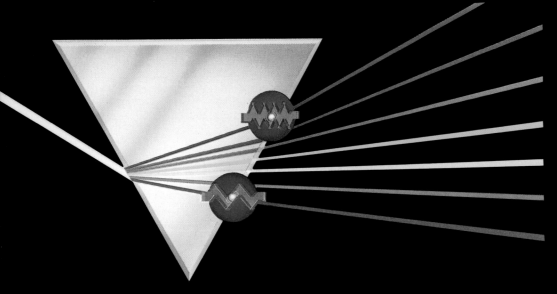

Dispersion—the spreading of white light into its constituent colors—occurs when a light beam passes through a glass lens or, as here, a prism. Violet light, whose photons have the highest frequency *(upper inset)*, interacts the most with the atoms of the glass; as a result, the violet light loses the most speed in the glass and is refracted at the steepest angle. By contrast, the low-frequency photons of red light *(lower inset)* interact comparatively weakly with the glass; the red light travels fastest through the prism and is bent the least by its passage.

Dispersion causes the lens defect known as chromatic aberration *(right)*. Because a simple convex lens bends violet light the most, that color comes to a focus nearest the lens; red light, bent the least, comes to the most distant focus, and the other colors reach their foci at points in between. Since light of only one color can be focused at a given distance from such a lens, stars and other light sources take on fuzzy, rainbow-colored halos *(far right)*.

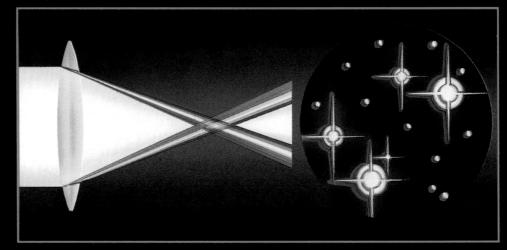

To cure chromatic aberration, telescope makers typically cement a concave lens to the back of the convex lens. Although the convex lens continues to converge violet light more sharply than red light, the concave lens offsets the aberration by introducing one of its own: It forces violet light to diverge more than red light, thus compensating for the chromatic aberration of its convex partner. This paired lens—called an achromatic doublet—focuses all colors at a single point, producing the sharper images seen at far right.

BATTLING THE RAINBOW EFFECT

White light, astronomers know, is nothing of the kind; rather, it is a mixture of all visible wavelengths, each of which produces a different color when isolated. The shorter wavelengths yield violet and blue light, those in the medium range produce green and yellow, and the longer ones create red and orange. When white light passes through a telescope lens, it tends to break apart into these constituent colors. This phenomenon—a complex side effect of refraction known as dispersion—must be corrected if the telescope is to deliver crisp images.

Dispersion occurs because the speed of light in a dense medium such as glass varies with the wavelength of the light. The photons of short-wavelength violet light, for example, oscillate at a very high frequency, causing them to interact often with the atoms of glass in a prism or lens. As a result, the violet light slows more in the glass than do other wavelengths and leaves the glass at a sharper angle than the rest. The photons of long-wavelength red light, by contrast, oscillate (and therefore interact) at much lower levels; the red light thus moves fastest through the glass and is deflected the least in transit.

This differential bending of light can wreak havoc on images formed by a simple lens. Violet light, angled the most, reaches a focus nearer the lens than does red light; the light of the intervening wavelengths, meanwhile, converges at locations between those two extremes. In consequence, the lens produces not one image but a series of single-color images, each focused at a slightly different distance from the lens (*opposite, center*). Since only one wavelength of light will be in focus at any given distance from the lens, an image made at a particular distance from the lens will be blurred by color halos.

This inherent lens defect, called chromatic aberration, can be remedied by attaching a second lens of differing shape and composition to the first. If the first lens is convex, the second lens will be concave. If the first is made of crown glass, the second will be made of flint glass, which has opposite dispersive properties. In such a hybrid lens, known as an achromatic doublet, each lens cancels out the chromatic defects of the other, bringing all colors to a focus at a single point. As shown below, achromatic doublets are key components of most modern refractors.

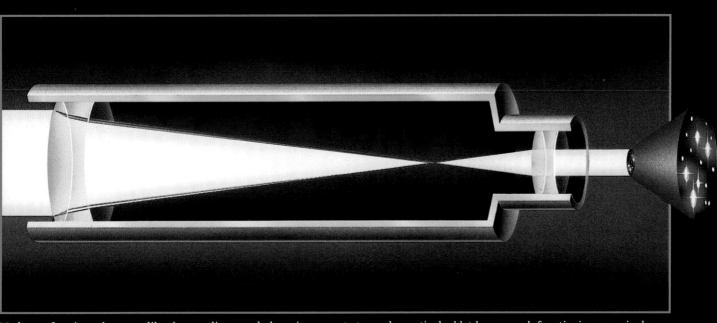

Modern refracting telescopes like the one diagramed above incorporate two achromatic doublet lenses, each functioning as a single lens. The doublet objective lens at left collects light and brings its various component colors to a sharp focus within the telescope tube. The doublet eyepiece magnifies the image produced by the objective, keeping the colors in uniformly sharp focus.

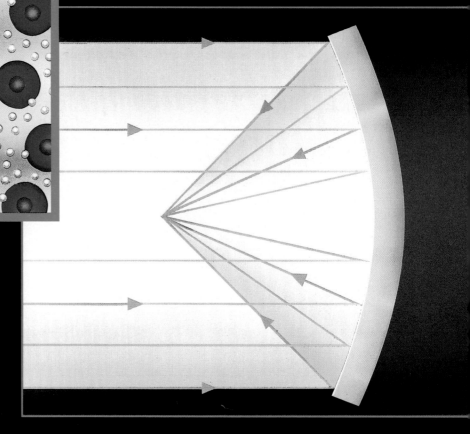

Photons of light striking an aluminum-coated mirror cause free electrons to oscillate and emit identical photons. The complex laws of electromagnetic theory dictate that the angle of reflection will equal the angle of incidence *(red arrows)*, whatever the wavelength of the photons.

A metal-coated parabolic mirror *(right)* reflects incoming parallel rays of light to a single focus. Because the mirror's outermost reflecting surfaces are slightly lower than those of a spherical mirror (one whose surface forms a section of a sphere), a parabolic reflector is free of spherical aberration, which is the result of different parts of a mirror having slightly different focal lengths.

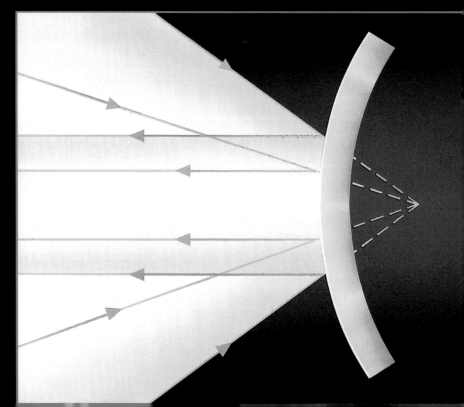

Although a convex mirror would diverge rather than converge incoming parallel light rays, it reflects converging beams of light as parallel rays *(right)*. This property enables the convex mirror to complement the action of large, concave light collectors in modern reflecting telescopes *(opposite, bottom)*.

REFLECTIONS IN A SILVER EYE

Although a refracting telescope can be cured of its chromatic aberration, the instrument's other design maladies are tougher to treat. A transparent glass lens can be supported only at its edges, for example, so the maximum practical diameter for a refracting objective lens is about forty inches; anything larger tends to sag out of shape from the force of its own weight. Turning molten glass into a bubble-free lens is also difficult and costly.

All such obstacles can be overcome by mirrors, which collect and focus light inside most modern telescopes. Operating with just one optical surface— usually, a thin

layer of aluminum or silver deposited on a curved disk of glass—a mirror can be supported from the back and thus can be much larger than a lens, yielding greater light-gathering area. Because only its surface interacts with light, the mirror's glass body need not be flawless. Best of all, a mirror reflects all wavelengths of light at a constant angle, eliminating the problem of chromatic aberration. As explained opposite, the optimum reflector shape is that of a paraboloid. A spherical mirror, though easier to grind, has steeper sides; these bring light from the edges to a focus nearer the mirror than light reflected from the center, a flaw known as spherical aberration.

A Newtonian telescope uses a side mounted eyepiece to keep the observer from blocking the light path. Incoming rays are reflected from a concave parabolic mirror, and the converging beam is intercepted by a flat mirror mounted at a 45-degree angle to the tube's centerline. The deflected beam reaches a focus near the side of the tube, where an eyepiece magnifies the image.

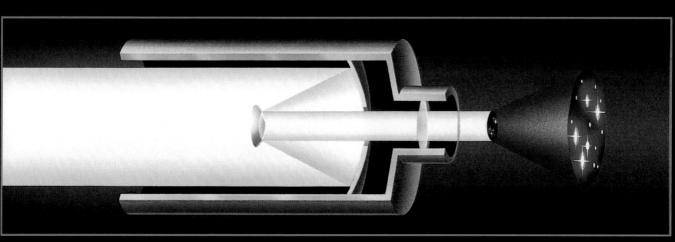

Most large modern telescopes are Cassegrain reflectors *(above),* with a parabolic primary mirror containing a central hole. Incoming parallel rays are reflected and converge on a convex secondary mirror, which restores their parallel orientation and channels them through the primary's hole to a focus behind it, where other lenses may magnify or correct the image even more.

Hall was no promoter, and his doublet languished for more than twenty years until a London optician by the name of John Dollond, learning of Hall's work, began to make such lenses. In 1757, Dollond built a 6-foot-long telescope with a doublet objective that produced a better image than a 120-foot-long refractor made with a single-element lens.

There was, however, a snag. While it was possible to make a crown-glass lens blank up to eight inches in diameter, the best that could be accomplished with flint glass was a disk about four inches across. For reasons no one could guess, larger blanks were marked by streaks and veins and were peppered throughout with little bubbles. Since light-gathering power decreases rapidly as the diameter of the lens shrinks, early doublet refractors with objectives no larger than four inches across sacrificed a measure of brightness for the sake of obtaining aberration-free images.

As the eighteenth century drew to a close, various scientific bodies in France offered prizes for substantial progress toward banishing streaks and bubbles from flint glass. The lure of such a reward may have spurred Pierre Guinand, a Swiss cabinetmaker, to try his luck. Some of his experiments went badly. On several occasions, for example, the vessel in which he melted the glass split open and the contents were lost. Once his roof went up in flames, and when the fire was doused, water fell onto the hot glass and ruined it. Persevering, he developed superior methods for mixing ingredients for glass, as well as for heating, stirring, and cooling the molten material, but for a time he was unable to cast satisfactory lens blanks larger than five inches. Then in 1805, Guinand discovered that the choice of rod used to stir the melt made a big difference. He had been blending the mixture with a wooden pole that became charred by the heat of the molten glass. When he turned instead to a rod made from fire clay, an extremely heat-resistant ceramic, a great many of the bubbles that had plagued the melt now rose quickly—almost magically—to the surface and burst. Moreover, the glass no longer developed streaks as it cooled.

Soon, Guinand was making lens blanks of flint glass as large as six inches in diameter and grinding them to create his own doublets. His lenses attracted attention from opticians and businessmen, one of whom established the glassworker in a new factory near Munich, Germany. There he concentrated on producing top-quality flint glass, delegating the work of grinding to a young German craftsman named Joseph von Fraunhofer.

A BAVARIAN GENIUS

Fraunhofer, an experimenter in the mold of Guinand, made four great contributions to telescopy during his association with the former cabinetmaker and thereafter. First, he put lens grinding on a scientific basis by devising precise mechanical and optical techniques—as well as the necessary measuring tools—to determine the deviation of a lens from the desired shape.

Astronomers locate celestial objects by marking their coordinates on the celestial sphere (above)—the apparent sphere of sky whose north pole, south pole, and equator are all extensions of their terrestrial counterparts. One coordinate, called right ascension (blue arrow), is similar to longitude and is measured eastward from the point (white dot) where the celestial equator crosses the ecliptic, the apparent path of the Sun across the sky. The other coordinate, called declination (red arrow), is the celestial equivalent of latitude and is measured in terms of degrees north or south of the celestial equator.

A telescope set in a so-called equatorial mount (right) is aligned so that its vertical, or polar, axis points to the north celestial pole in the Northern Hemisphere, to the south pole in the Southern Hemisphere. Rotating the telescope around this axis (blue arrow) shifts the instrument east or west. Rotating it around its declination axis (red arrow) effects north-south movement. Once a star is in focus, the declination axis remains fixed and the telescope rotates around the polar axis to keep pace with the star's apparent westward movement as Earth rotates.

Fraunhofer performed a similar service for measuring the refractive powers of a given piece of glass. An optician needed accurate refraction information in order to grind a lens to prescription. Because glass bends light of longer wavelengths less than light of shorter wavelengths, refraction measurements depend on the color of the light source, for which there was no standard among opticians of Fraunhofer's era. His contribution began with the discovery that the metal sodium, when heated in a flame, always produces an intense yellow light. Adopting the glow of sodium as the means to gauge the refractive power of glass, Fraunhofer suggested, would assure that such measurements would be consistent and comparable to one another.

A third contribution was to improve the performance of doublet objectives by making each element asymmetrical. That is, he ground the two surfaces of both the convex and the concave lenses with different degrees of curvature rather than identical ones, as had been the practice. This innovation gave a lens designer additional flexibility in calculating the best prescription for an objective.

Rounding out Fraunhofer's quartet of contributions was a vastly improved version of a support for telescopes called the equatorial mount. Known in its essential characteristics to Tycho Brahe and his successors, the equatorial mount permitted a telescope to follow a particular star across the sky simply by being turned on one axis rather than two *(left)*. Fraunhofer built his mount largely of steel for rigidity. Furthermore, he supplied it with counterweights to balance the mass of the telescope, which made aiming the heavy instrument almost effortless. Fraunhofer's equatorial mount and subsequent variations on the theme were invaluable for providing stable, steerable support for the larger-diameter, heavier refractors that were to come.

Between 1820 and 1824, Fraunhofer helped to produce the largest refracting telescope yet built. By then, size in telescopes had generally come to refer to the diameter of the objective rather than the instrument's length. With this telescope, erected at the Dorpat Observatory in Estonia, astronomers peered at the universe through a doublet lens that was nine and one-half inches across. The telescope was set in an equatorial mount equipped with a drive reminiscent of the works inside a cuckoo clock; it was mobilized by the action of a weight slowly descending on a chain. Using the Dorpat refractor, the great German astronomer Friedrich Georg Wilhelm von Struve conducted a fruitful search of the cosmos for double stars. The Fraunhofer-designed objective allowed Struve to distinguish the individual stars of thousands of very closely spaced pairs that earlier had been thought to be single stars.

The Dorpat telescope's 9.5-inch objective was an important milestone in the development of the refractor. Yet other instruments would surpass it. One to do so was a 15-inch doublet telescope installed at Harvard College in the late 1840s. The lens, acquired in Germany, made the Harvard refractor one of the best such instruments of its day. Among its more important contributions was the influence it exerted on American portrait painter Alvan Clark, who chanced to look through the new Harvard refractor shortly after its inauguration. Upon ascertaining that the objective lens had cost $12,000, an immense sum at the time, Clark resolved to learn to grind telescope lenses and to make his fortune at it. He experimented by regrinding discarded lenses to new focal lengths, and he succeeded in making refractor objectives of very high quality. He then set up a lens-making business with his two sons Alvan and George.

Benefiting from European advances in glassmaking, Clark and Sons produced objective lenses for six telescopes, each of which was the world's largest refractor at the time it was installed. In 1872, the firm completed a 26-inch lens for a telescope at the U.S. Naval Observatory in Washington, D.C., with which the astronomer Asaph Hall discovered the moons of Mars in 1877. Fifteen years after the Navy's lens was ground, a 36-inch Clark objective was placed into service at Lick Observatory near San Jose, California. The Lick refractor was used to conduct an extensive study of the radial velocities of stars—that is, their velocities toward or away from the Solar System. From this survey, which continued for more than half a century, astronomers plotted the motions of stars and measured the velocity of the Sun with respect to its neighbors.

The greatest of the Clark refractors, the 40-inch telescope of the Yerkes Observatory in Williams Bay, Wisconsin, was finished in 1895. The Yerkes refractor was the biggest ever made, and it is likely to remain so. Even with modern glassmaking technology, producing a flawless lens blank wider than the one used for the Yerkes objective would be a costly undertaking with little chance of success. Furthermore, a larger lens would droop unacceptably under its own weight, thereby distorting and blurring images. And since a lens must let light pass, it can only be supported along its edge, making the sag impossible to correct. To peer deeper into space with greater acuity, astronomers turned to the antithesis of lenses—mirrors.

REFLECTIONS OF THE HEAVENS

In antiquity, mirrors were made from polished brass and other metals and, according to the Roman naturalist Pliny, from glass covered with a soft metal such as tin or silver. Mirrors held the same fascination as lenses for early investigators in the field of optics. The Arab physicist Alhazen paralleled his work in refraction—as did Kepler—with experiments on the reflecting properties of curved metal surfaces. And in Britain during the late sixteenth century, a few experimenters tried unsuccessfully to make a spyglass-like instrument from crude concave and convex mirrors. But no significant

Lens maker Alvan Clark *(above, left)* and his assistant Charles Lundin administer a final touch-up to the 40-inch lens for the Yerkes Observatory at Williams Bay, Wisconsin. In all, Clark and his sons Alvan and George built refractors for six telescopes—each the largest in the world at the time it was installed.

progress was made in the field of reflecting telescopes until sixty years after Galileo's earliest refractors.

Credit for the first reflecting telescope worthy of the name goes to Isaac Newton. In the course of his experiments in the mid-1660s with prisms and lenses, Newton acquired the erroneous notion that all types of glass bend light equally, and he logically concluded that refractors could never be made free of chromatic aberration. His solution to the problem was to avoid refraction altogether by using a mirror made of polished speculum, a metal alloy composed mostly of copper and tin. Reflected from the surface of the mirror— and thus passed through nothing refractive—the light would not be split into its spectrum, and the characteristic halos of chromatic aberration would not form.

Newton brought all the necessary skills to the mirror-making job. Not only did he draw up the optical design of the telescope, but he also mixed his own highly reflective mirror alloy, using a recipe of six parts copper, two parts tin, and one part arsenic. He cast mirror blanks from the molten speculum metal; then, using techniques of his own invention, he ground and polished them with grains of abrasive mixed in putty. Newton designed a reflector consisting of a concave primary mirror—analogous to the refracting telescope's objective lens—that collected the light and focused it in a converging beam toward a small, flat secondary mirror placed on the centerline of the telescope tube, near the top. The secondary mirror intercepted the light from the primary and reflected it through a hole in the side of the tube, where Newton could view the image with an eyepiece. Although the diameter of the primary mirror in this first reflector was only about one and one-third inches, it worked well. In early 1669, Newton told a friend that the reflector magnified "about 40 times in Diameter" and that "I have seen with it Jupiter distinctly round and his Satellites."

An error contributed greatly to the telescope's success. Newton had planned to give his primary mirror a spherical curve, even though he was aware that by doing so he would introduce the spherical aberration that plagued lenses of the era. In the event, however, his polishing technique was flawed. It produced a parabolically curved surface that, like a hyperbolic one, focuses light originating near the optical axis at a single point. Like parabolic or hyperbolic lenses, mirrors having these kinds of curvatures focused off-center objects much less sharply than a spherical mirror.

In late 1671, he allowed his Cambridge professor, Isaac Barrow, to take the new reflector to London, where Christopher Wren and other prominent thinkers demonstrated it to Charles II. The instrument created a sensation. The Royal Society placed an order for two reflecting telescopes with a London optician and elected young Newton to their elite membership.

A SLOW START

Although the greatest telescopes in the world today are all reflectors that can be traced to Newton's invention, the development of the technology was far from smooth. For example, a mirror made of speculum metal, with its high copper content, tarnished rapidly. Within a year or two—or considerably less time, depending on climate—a mirror became so dim that the faintest stars disappeared from view. To remove the tarnish, the mirror had to be repolished, a time-consuming process that risked spoiling the shape. For this reason and others, the mirror's lack of chromatic aberrations was not enough to win astronomers away from lenses, and progress in reflecting telescopes poked along for decades.

But in the early 1720s, when long-focal-length objective lenses were beginning to turn refracting telescopes into ungainly monsters, the idea of a reflecting telescope was reexamined by Englishman John Hadley. He constructed an instrument with a speculum mirror six inches in diameter. In competition against a tubeless refractor 123 feet long, the reflector acquitted itself well in its ability to resolve faint stars. Being only five feet long, the reflector was much easier to aim, and it had the further advantage that, because the mirrors were enclosed in a tube, extraneous light could not wash out the image.

More important than Hadley's telescope to a revival of interest in reflectors was a method he introduced for optically inspecting the surface of a concave mirror as it was polished. Hadley illuminated the mirror with light shining through a pinhole placed above it at the center of curvature—that is, the center of the sphere of which the reflecting surface was a part. Looking at the mirror, he could plainly see patterns in the reflected light that indicated where more grinding and polishing were needed. Requiring no complicated instruments or gauges, Hadley's technique not only simplified the fabrication of telescope mirrors but also eliminated much of the risk to the shapes of mirrors inherent in repolishing them.

The premier mirror maker of the day was William Herschel, a German musician who moved to England as a young man and became an amateur astronomer in middle age. Because he could not afford to buy a large telescope

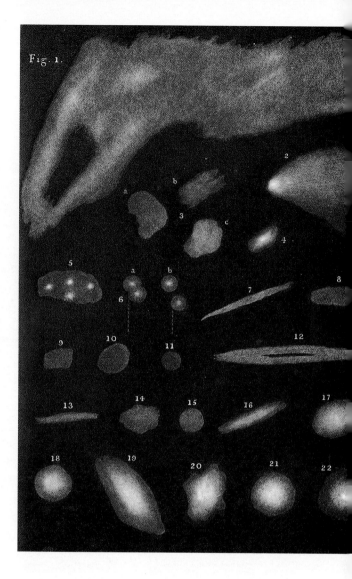

Fig. 1.

Between 1773 and 1789, William Herschel built a series of large reflecting telescopes and used them to make the most detailed observations of his time, producing sketches such as this array of nebulae arranged by size, shape, and brightness. With his powerful telescopes, Herschel was able to resolve the stars in these fuzzy objects, leading him to subscribe, albeit temporarily, to an early theory that the nebulae were what he termed island universes—that is, separate galaxies.

from an optician, he taught himself to make mirrors and was soon building the best telescopes the world had seen, unsurpassed in their ability to resolve fine detail and reveal faint stars.

Herschel kept as trade secrets the details of how he produced mirrors but claimed that his training on the violin gave him skill at delicate polishing strokes. A perfectionist, he would test a mirror again and again by taking it from his optical workshop to the telescope, checking its performance, then returning it to the shop for more polishing. His first successful telescope, patterned on Newton's design, had a mirror six and a half inches in diameter and a tube seven feet long. While peering through this instrument one night in 1781, he spotted an object that was perceptibly larger than the stars around it. Convinced by the fuzziness of the image that he had found a comet, he changed eyepieces to increase magnification. Only after extended observation of his find, however, did Herschel realize that he had come upon not a comet but a new planet. He named it Georgium Sidus—George's Luminary—in honor of King George III; today, it is known as Uranus.

Following this momentous discovery, the king asked that Herschel's impressive telescope be taken to the Royal Observatory at Greenwich and tested against the instrument there. "We have compared our telescopes together," Herschel wrote home to his sister Caroline, "and mine was found very superior." George III, an astronomy enthusiast himself, granted Herschel an annual salary of £200.

During his studies of the "space-penetrating power" of telescopes, as he called their light-gathering ability, Herschel noticed that a mirror returned only part of the light that fell on it, absorbing the rest. In the layout for a Newtonian telescope, the light was reflected twice and dimmed to slightly more than half its original brilliance. So Herschel resolved to build telescopes having a single mirror. To eliminate the secondary mirror, he set the primary at a slight tilt in the base of the telescope so that it focused light directly into an eyepiece mounted just inside the top of the tube.

A MAN OBSESSED

Making telescopes by day and observing with them by night, Herschel came to neglect his music pupils, clothes, and personal appearance. Sometimes the work of figuring a mirror engrossed him for sixteen hours at a stretch. Recalling such episodes, Caroline told of being "constantly obliged to feed him by putting the victuals by bits into his mouth." Herschel made fifty or so telescopes between 1782 and 1785. Some he built for himself; others he sold to supplement his stipend from the king. As he gained experience, he made several primary mirrors for most of his reflectors. When the surface of one mirror tarnished, he simply replaced it with another.

Before Herschel's telescopes, astronomy was limited largely to the study of the planets; stars served chiefly as reference points for charting planetary motions. But Herschel's superb reflectors permitted him to study many more double stars than had previously been visible. He also cataloged thousands of star clusters and objects called nebulae, thought to be clouds of interstellar gas and dust. Some nebulae were just that. Others, however, viewed with a 19-inch reflector Herschel built, proved to contain vast numbers of stars. He concluded that "nebulae" of this kind are in fact galaxies, like the Milky Way—the only galaxy known at the time.

To gain a better idea of the dimensions of the Milky Way itself, Herschel invented the technique of star gauging. He counted all the stars that he could see through a given telescope aimed successively at many points in the sky. Herschel called out the star counts to Caroline, who recorded them in a log. He theorized that areas with a greater density of stars corresponded to directions in which the Milky Way extended farther along the line of sight. After much work, he concluded that the Milky Way is flat like a grindstone and that the Solar System is not at the center.

To chart faint stars in the Milky Way and thereby (as he thought) see through the full extent of the galaxy, Herschel needed to build telescopes with still more light-gathering power. In 1785, he began work on a reflector with a 48-inch mirror. The telescope, which was financed by a grant of £4,000 from George III, had a tube fashioned from sheet iron that was almost 40 feet in length; a wooden structure was erected to support and steer it. Three primary mirrors were cast. The first was serviceable, although it was thinner at the center than intended. The second cracked. The third was close to perfect. On August 27, 1789, it was hoisted into position with a crane. On the night following the telescope's completion, Herschel, peering into the new reflector, discovered Enceladus, the sixth moon of Saturn. On September 17, he discovered the seventh, Mimas.

Despite early success, the telescope was doomed to failure by the size of its mirror; it was so huge that it sagged under its own weight, making clear images difficult to obtain. It was also inconvenient to use, with at least two assistants being required to help point it and record observations. And tarnishing took its toll: By 1801, Herschel found that the mirror was "much injured by time." In 1815, he abandoned the telescope.

The largest reflector built with a metal mirror—a 72-inch monster—entered service at Birr Castle, Ireland, thirty years later. Known as the Leviathan of Parsonstown, the huge Newtonian reflector was commissioned by William Parsons, the third earl of Rosse and an avid amateur astronomer. Its four-ton primary mirror enabled Lord Rosse to discern the spiral structure of some nearby galaxies and the filamentary shape of the Crab nebula, which he named. However, the Leviathan could rarely be used to full advantage because the sky in Parsonstown was so often overcast, and the telescope gradually fell into disuse.

THE DECLINE OF SPECULUM
A point of diminishing returns had now been reached with speculum metal mirrors. Not only did tarnishing rapidly weaken their light-gathering power, but also changes in temperature would cause them to expand and contract

considerably. These stresses could alter the figure of the mirror, diminishing its resolution. Lord Rosse had rigged levers behind his mirror to compensate for such distortion, but the system did not work reliably. On some nights, he could adjust the mirror to produce extraordinarily sharp images; on others, the telescope's performance was a disappointment.

For these reasons, telescopes with lenses remained the instrument of choice among professional astronomers during the second half of the nineteenth century, even though the solution to all the reflector's problems was at hand: Around 1850, scientists in Europe had discovered how to deposit a layer of silver on a glass surface—even a concave one, as needed for the primary mirror of a telescope—by covering it with a solution of silver nitrate and sugar. After a time, fine granules of silver precipitated out of the solution and stuck to the glass. This reaction produced a thin, uniform, and highly reflective deposit.

The advantages of silver over speculum metal were manifold. The shiny metal returned a third more light and tarnished less rapidly than the alloy. Removing the discoloration was a relatively simple matter of dissolving the old silver and depositing a new coating, a process that did not demand reworking the concave figure of the mirror. Glass was cheaper than speculum metal—and lighter, too, so that a large mirror sagged less and could be more easily mounted and supported. Moreover, glass was less susceptible than metal to the distortions that could be caused by temperature changes.

Most prominent among those who experimented with glass telescope mirrors was the French physicist Jean Foucault. He made several small Newtonian reflectors during the late 1850s, the largest having a mirror less than eight inches in diameter. Noting that gravity caused even these mirrors to droop ever so slightly, Foucault remedied the situation by backing the mirror with a rubber bladder. While looking into the telescope, he could inflate or deflate the bladder to subtly change the shape of the mirror and thereby achieve resolution that approached the instrument's theoretical limit.

This boundary of telescope performance had been established in the 1830s by British astronomer George Airy, who discovered that the best resolution obtainable from any mirror depends on the ratio of its diameter to the wavelength of light falling on it. Thus, the wider the mirror, the better

The remnants of William Herschel's 40-foot-long telescope lie in state in the garden of his observatory in Slough, England. With this instrument's 48-inch mirror, Herschel found Saturn's sixth and seventh moons in August and September of 1789. Because the copper mirror tarnished quickly and the telescope itself was difficult to operate, Herschel used it only rarely, and abandoned it after 1815.

its resolution should be. In practice, however, the rule fails for mirrors larger than about eight inches in diameter. Distortion of light waves by Earth's atmosphere nullifies the potential gain in resolution from wider mirrors. Their chief value lies in their ability to gather more light, thereby revealing the presence of heavenly bodies invisible to lesser optics.

In 1864, Foucault completed a mirror more than 30 inches wide. Originally installed at the Imperial Observatory in Paris, it was soon moved to Marseilles. With more than ten times the light-gathering power of Foucault's 8-inch reflector, the instrument enabled astronomer Edouard Stephan to catalog hundreds of nebulae. Despite this and other successes of the 30-inch mirror, Foucault's reflectors—and larger ones that followed before the turn of the century—were for the most part unproductive. Poor telescope design and underengineered mounts were at fault more often than the mirrors. A few such telescopes met their designers' expectations for image quality and utility, notably a French reflector of 47 inches made in 1877 and a 36-inch instrument installed at California's Lick Observatory in 1895. But a full realization of the potential of such devices awaited the energy and dedication of American astronomer George Ellery Hale.

REFLECTORS REASCENDANT
Born in 1868, Hale was the son of a wealthy Chicago businessman who manufactured hydraulically actuated elevators. He studied physics as an undergraduate at the Massachusetts Institute of Technology, where he demonstrated a knack for instrumentation. He also proved to be a talented organizer who possessed remarkable powers of persuasion. Here was a man who could talk turkey with industrial tycoons and at the same time make sense to scientists.

These skills, rarely combined in one individual, enabled Hale to become the dynamo behind some of the most important telescopes ever built. The Yerkes Observatory, home of the world's largest refractor, was his idea. With money that he solicited from a Chicago streetcar magnate while still in his twenties, he had been the one to commission Alvan Clark and his sons to grind the 40-inch lens that made the telescope there famous.

Realizing that refractors had no further potential, Hale persuaded his father in 1896 to buy him a blank for a 60-inch mirror. The Carnegie Foundation, a philanthropic organization started by business tycoon Andrew Carnegie, put up the money for a telescope equipped with the mirror. To provide a home for the instrument, Hale founded the Mount Wilson Observatory in 1904, using his own funds as seed money. By the year 1908, the 60-inch reflector was at work.

Mount Wilson, in the San Bernardino Mountains east of Los Angeles, was an excellent site for the telescope. Situated more than a mile above sea level, the facility rose above much of the star-obscuring dust and smoke in Earth's atmosphere—and higher than any other observatory of its day. By reducing atmospheric distortion, the altitude contributed to the telescope's unprecedented performance. It was, for example, able to resolve stars in the Andromeda nebula that, through the Yerkes 40-inch refractor, had appeared as so much cosmic dust.

In 1907, Hale persuaded Los Angeles hardware entrepreneur John D. Hooker to pay for a 100-inch reflector that would join the 60-inch reflector on Mount Wilson and bear its benefactor's name. When funds for the Hooker Telescope ran out, Hale talked the Carnegie philanthropists into making up the difference. By 1918, the Hooker Telescope was peering into the universe, helping to establish—far more accurately than Herschel had been able to—the size of the Milky Way and the number of stars it contains.

Hale believed that the future of astronomy lay in the new field of astrophysics—the application of physical laws studied in the laboratory to celestial objects investigated in the observatory. This discipline uses spectroscopy to study stars through analysis of the electromagnetic radiation produced by the internal nuclear reactions that power them. All elements emit distinctive wavelengths of light—their spectra—by which they can be identified. Furthermore, the phenomenon known as the Doppler effect makes light appear to be of slightly shorter wavelengths when it comes from an object moving toward the observer and of slightly longer wavelengths when it is receding. Since the magnitude of this Doppler shift varies with speed, spectroscopy provides the key to charting the courses of stars and galaxies as they transit the cosmos.

Hale made sure that each of the Mount Wilson reflectors was equipped with a spectrograph—the special instrument needed to make such studies. The device was attached at the base of the telescope, behind the primary mirror. Such a setup was possible only because Hale had abandoned the Newtonian design common to most earlier reflectors, using instead an arrangement of mirrors proposed in 1672 by Guillaume Cassegrain, an astronomically inclined Frenchman of uncertain occupation. (He is thought to have been either a sculptor or a university professor.) Cassegrain's scheme substituted a convex mirror for the flat secondary reflector of Newton's telescope. This mirror, instead of directing the image to the side of the telescope tube, focused it back through a hole in the center of the primary mirror *(page 39)*. The design, though difficult to execute because it required the grinding of two curved mirrors, more than halved the length of the tube required for a reflector of a given focal length.

Mount Wilson astronomer Edwin P. Hubble and coworker Milton Humason made good use of the spectrographs hitched to the two reflectors there. After devoting six years to analyzing Humason's many spectrograms of faint galaxies, Hubble announced in 1929 that all but a few nearby galaxies are receding from the Milky Way at rates proportional to their distance. That is, the more distant the galaxy, the faster it is moving away. The discovery revolutionized cosmology. Heretofore, most astronomers had thought the universe to be static—neither increasing nor decreasing in its dimensions. The Mount Wilson astronomers proved that it is expanding.

By this time, Hale had officially retired. For much of his life, he had suffered from painful headaches, vivid hallucinations, and nervous breakdowns. These infirmities were aggravated by responsibilities that reached far beyond the narrow world of astronomy. At his urging, for example, the National Research Council had been formed to advise President Woodrow Wilson's administration on scientific matters during World War I. He had also joined the board of trustees of Throop Polytechnic Institute, a 550-student engineering college in Pasadena, California. In 1920, acting on a Hale proposal to make Throop the MIT of the west, the board renamed the school the California Institute of Technology.

A MAN FATIGUED

At age fifty-four, heeding the advice of his physician, Hale went abroad for rest and recuperation. But upon returning home, he set out on yet another campaign: to provide astronomers with a more powerful telescope than the Hooker. Abandoning a concept for a 300-inch reflector as impractical (the primary mirror for such a behemoth would have been too large to ship from manufacturer to observatory, he thought), Hale lowered his sights and settled on a 200-inch mirror. In an article written for the April 28, 1928, issue of *Harper's Weekly*, he once again framed an eloquent appeal to the worlds of money and science, acknowledging the "princes and potentates, political or industrial" who had financed his earlier telescopes and bemoaning the fact that "starlight is falling on every square mile of the earth's surface, and the best we can do at present is to gather up and concentrate the rays that strike an area 100 inches in diameter." Hale's plea was heard by the Rockefeller Foundation, which agreed to fund not only the telescope but also an observatory on California's Mount Palomar to shelter it. Altogether, a little more than $6.5 million would be spent.

Corning Glass Works of Corning, New York, produced the blank for the mirror of Pyrex, the trademark name of a new type of glass that was much less prone to change shape in response to variations in temperature than were earlier formulations. For stiffness, the casting was made two feet thick, but not of solid glass. A ribbed back *(page 60)* kept the weight of the casting to 29,000 pounds. The ribs would also enable the finished mirror to match the temperature of the night air more rapidly than could a solid slab of glass —and thus diminish relatively quickly the disruptive convection currents rising from its surface.

The blank was transported by rail from New York to Pasadena, where Caltech opticians ground and polished the glass surface into a nearly perfect parabola. Begun in 1936 and interrupted for several years by World War II, the process was not completed until 1947. By that time, Hale was nine years dead, victim of a variety of ailments, and Max Mason—a professor of mathematics and physics, as well as a director of the Rockefeller Foundation—had taken over the project.

The precipitation process for applying a reflective coating to a polished blank had been made obsolete in 1932 by the invention of a superior technique. The brainchild of American physicist John Strong, the new method called for placing the blank in a vacuum chamber that also contained thin wires connected to a powerful source of electricity. When sufficient current was passed through the wires, they exploded, vaporizing onto the glass a microscopically thin layer of metal from the wires that exactly mimicked the carefully figured surface of the mirror. No polishing or buffing was necessary. Instead of silver for the wires, Strong used aluminum to coat the 200-inch mirror, the lighter metal being more reflective and tarnishing less readily.

The finished mirror was hauled to the summit of Palomar Mountain aboard a flatbed truck, then hoisted into its huge equatorial mount three days before Christmas, 1947. The completed instrument, weighing some 530 tons and able to see twice as far into space as the 100-inch reflector at Mount Wilson, remains a colossal achievement of optical engineering *(pages 52-63)*.

At a dedication ceremony six months later, the new telescope, which until then had been referred to simply as "the 200-inch," was named after the driving force behind it—George Ellery Hale. After an extensive shakedown period, the Hale reflector went into regular service in 1949. Among numerous other contributions, the telescope played a part nearly fifteen years later in the enumeration of the quasars that currently define the visible edge of the universe.

Few imagined, when the Hale Telescope made its debut, that it would ever be surpassed in its light-gathering ability: The mirror required for a superior reflector would simply be too massive, the time to shape it too extensive, the structure to support it too expensive. But observational astronomers had other tricks up their sleeve. With the traditional avenue of larger objectives blocked, they sought to enhance their view by equipping telescopes with electronic eyes that can wring every last photon of information from the light show of the cosmos.

The driving force behind some of the world's most important telescopes, American astronomer George Ellery Hale scouts the landscape from California's Mount Wilson in 1903. As founder and first director of the Mount Wilson Observatory, he oversaw the building of its 60-inch and 100-inch reflectors, as well as the great 200-inch mirror at Palomar Observatory that bears his name.

THE TWO HVNDRED INCH DOME

AN ENDURING GIANT

Sheltered from wind and weather within its mountaintop dome, the Hale Telescope is a marvel of design and engineering that has remained the standard in the field since it began regular operation in 1949. For nearly three decades, its 200-inch primary mirror ranked as the largest and most powerful in the world, capable of detecting a single candle at a distance of 15,000 miles. Even the mammoth 263-inch mirror at Mount Pastukhov in the Soviet Union, which saw first light in 1976, has never matched the resolution afforded by the telescope on Palomar Mountain.

Over the years, the Hale has proved itself remarkably versatile as well. In addition to the primary mirror, it possesses an array of smaller mirrors that can deflect light to one of four alternate focal points, each equipped with instruments such as cameras and spectrographs. Although photographic plates were once the principal means of recording light, the Hale has been continually updated with a succession of state-of-the-art devices, including digital cameras that substitute computer chips for film *(pages 92-93)*.

Building the Hale took more than two decades, with time out for World War II. Early in the process, however, the team of astronomers and engineers knew what their handiwork would eventually look like. With nothing but mechanical blueprints to work from, scientific illustrator Russell W. Porter brought the telescope to life in a series of drawings done during the 1930s and 1940s. Porter's vision of the future is as accurate as it is beautiful, from his portrait of the 135-foot-tall observatory *(left)* to the detailed renderings, shown on the following pages, of the telescope's inner works.

ANATOMY OF THE PALOMAR BEHEMOTH

Resting at the bottom of an open-sided tube (shown here in its upright, parked position), the Hale's primary mirror is supported by a yoke harnessed to an arrangement of gears that allows the telescope to rotate from east to west and from north to south. After light rays have traveled through the telescope tube to the primary mirror, they can be directed to converge at a given focal point, depending on the brightness of the object under study and the aspect of its light being examined.

Of the Hale's four focal points, the three shown at right are used regularly. Prime focus, the point at which light reflected by the primary mirror converges, is located 660 inches above the mirror and is housed in a cagelike workspace large enough to accommodate an astronomer. A convex mirror swung into the path of light bouncing off the primary mirror diverts the light rays through a hole in the center of the primary to converge at the Cassegrain focus. With a focal length of 3,200 inches, the Cassegrain focus produces images that are about five times larger than those at prime focus. For example, an image of the Moon at prime focus has a diameter spanning approximately six inches; at the Cassegrain focus the diameter is three feet. Since the light is spread over a larger area, the image is not as bright as it is at prime focus, but it reveals more details.

Yet another focus, located in a room on the level below the telescope, has an even longer focal length—6,000 inches, resulting in images nearly nine times larger than those at prime focus. Light follows a crooked path to this focal point, earning it the name of coudé—from the French for "elbow." To reach the coudé focus, light reflected from the primary mirror hits the convex coudé mirror below prime focus, then bounces off a series of flat mirrors. When the Cassegrain and coudé mirrors are not in use, they are folded into recesses in the walls below prime focus.

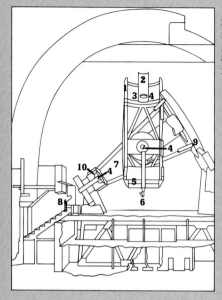

1: telescope tube. 2: prime focus. 3: Cassegrain mirror. 4: coudé mirror. 5: primary mirror. 6: Cassegrain focus. 7: tubular girders. 8: coudé focus. 9: horseshoe bearing. 10: south polar axis bearing.

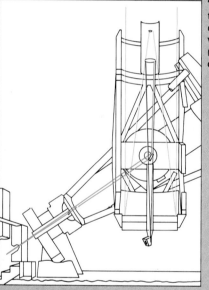

Through judicious use of mirrors, the Hale can route light to several different focal points, three of which are shown here: prime focus *(yellow)*, Cassegrain focus *(red)*, and coudé focus *(blue)*.

LIGHT PATH TO PRIME FOCUS f 3.3
" " CASSEGRAIN " f 16
" " COUDE " f 30

APPROXIMATE SCALE

R.W.PORTER '38

THE TWO HVNDRED INCH TELESCOPE~

Controlling a Well-Oiled Machine

The lattice of steel I-beams that forms the telescope tube weighs 140 tons, yet there is no more than a hundredth of an inch difference in flexure at each end when the tube is moved from its resting vertical position into its slanting work posture. As a result, the Cassegrain and coudé mirrors at the tube's upper end remain in near-perfect alignment with the primary mirror fifty-five feet below.

The telescope tube hangs between two tubular girders, each ten feet in diameter. The girders are supported in turn by two bearings—the so-called horseshoe bearing at the northern end and the south polar axis bearing at the southern end. The bearings are cradled by oil pads that allow the yoke and the telescope tube to be moved with extraordinary ease. A pumping system forces oil between the bearings and the pads at a pressure of 300 pounds per square inch. Floating on a layer of oil only a few thousandths of an inch thick, the bearings reduce friction so effectively that the combined 530-ton weight of the telescope tube and the yoke will respond to sustained pressure from one finger. Because of its sensitive response to any force, the telescope requires dampers to prevent light gusts of wind from shifting it about.

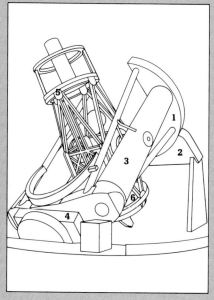

1: horseshoe bearing. 2: oil pad housing. 3: girders. 4: south polar axis bearing housing. 5: telescope tube. 6: primary mirror and protective dust cover.

A view of the horseshoe bearing's curving underside shows two of the four twenty-eight-inch-square oil pads with their supply lines and pressure gauges. The lines feeding oil to the pads' surfaces are continuously monitored, and power to drive the telescope is cut off automatically if oil pressure drops to an unsafe level.

THE · TWO · HVNDRED · INCH ~
TELESCOPE · LOOKING · NORTHWEST

Keeping Pace with the Stars

As the Hale points skyward, stars in its field of vision seem to drift to the west because of Earth's rotation on its axis. To compensate for this apparent motion and to maintain a fix on a celestial target, the telescope tube rotates steadily from east to west around its polar axis, so called because it is parallel to Earth's axis and points to the celestial north pole.

At the south, or lower, end of the polar axis is the right ascension drive, which regulates the east-west rotation. The drive's name alludes to one of the coordinates astronomers use to pinpoint an object's position in the sky. Just as locations on Earth's surface are denoted by the intersections of lines of latitude and longitude—imaginary lines that, respectively, parallel the equator and stretch from pole to pole—locations in the heavens are similarly identified by lines of declination and right ascension, paralleling the celestial equator and running between the celestial poles.

The right ascension drive has two gears, each fourteen feet in diameter and edged with 720 teeth. The slewing gear, with a two-horsepower motor, controls the telescope's large, rapid east-west movements—swinging it into viewing position when the night's work begins or shifting it from one target object to another. The tracking gear has two small motors of about one-twelfth horsepower each. One is computer controlled and is used to point the telescope; the other rotates the telescope ever so slowly on its axis to keep the target object precisely in view.

In addition to the east-west movement, astronomers can rotate the Hale north-south around its declination axis—by means of a fourteen-foot-wide gear in the west declination trunnion—in order to aim the Hale at virtually any point in the northern sky. (The horseshoe bearing prevents the telescope from seeing directly below the north celestial pole.) One of the gear's two motors is computer controlled to govern precise positioning; the other motor controls slewing. The computers for both the right ascension drive and the west declination trunnion allow the telescope to track a comet or other object whose movement—unlike the nightly progression of the stars—is real rather than apparent.

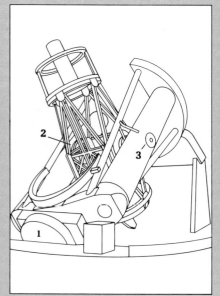

1: right ascension drive. 2: west declination trunnion. 3: east declination trunnion.

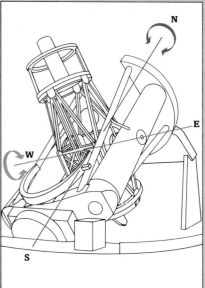

Rotation around the declination axis (blue line) moves the telescope in a north-south direction (blue arrow), while rotation around the polar axis (red line) shifts it from east to west (red arrow).

TWO HVNDRED INCH MIRROR
SVPPORT & COVER

CASSEGRAIN
FOCUS
ƒ 16

APPROXIMATE SCALE
0   1   2   3   4   5   6 FT

When the mirror is tilted on edge *(far left)*,
weights *(red)* that pivot on a system of bearings
force one end of the inner shaft *(green)* down,
thereby forcing the so-called mirror collar
(green) and the other end of the inner collar
(blue) up and applying upward thrust to the
walls of the mirror's recesses. When the
mirror is faceup for storage *(near left)*, the
thrust of the weights is transmitted directly to
the central shaft and on to the mirror collar
and the mirror.

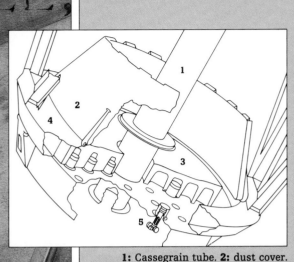

1: Cassegrain tube. 2: dust cover.
3: primary mirror. 4: mirror cell.
5: mirror supports.

FIGHTING THE FORCE OF GRAVITY

For sharp, distortion-free images, the surface of a telescope's primary mirror must reflect all the light that arrives from the target object to a common focal point. In addition to near-perfect smoothness—bumps would deflect light to points other than the one desired—a near-perfect configuration is also critical. In a mirror the size of the 200-inch Hale, which weighs fourteen and a half tons and is nearly two feet thick, the deadliest foe to maintaining the necessary paraboloid shape is gravity. As the mirror makes its nightly rounds, shifting from its face-up resting position to various degrees of tilt, it is prone to warp and sag under its own weight.

To counteract gravity's pull, the Hale mirror rests in a welded-steel structure, called the mirror cell, that secures the mirror to the bottom of the telescope tube and consists of two round, inch-thick plates. The upper plate has thirty-six holes that are aligned with thirty-six recesses in the mirror's underside. Through the recesses, a sensitive mechanism of weights, levers, and shafts *(diagrams, bottom left)* supports the mirror and delivers upward thrust, preventing the mirror from distorting, no matter how it is tilted.

A heavy, metal dust cover *(diagram, above)* safeguards the mirror against dirt and falling objects, as well as insulating it from temperature changes during the day. Folding like the petals of a flower, the dust cover's sections also function like a camera's iris diaphragm.

Teaching an Old Telescope New Tricks

Although it is no longer the world's largest telescope, the Hale remains a state-of-the-art instrument. Over the years, new detectors have extended its range beyond the narrow range of visible light into the shorter-wavelength ultraviolet segment of the spectrum and, in the other direction, the longer wavelengths of infrared energy, opening up such once-hidden vistas as the center of the Milky Way and the inner workings of stellar nurseries.

Modernization has changed the kind of work astronomers do with the Hale and where they do it. As illustrated here, for example, the prime focus cage near the top of the telescope tube *(below)* was the site of near-nightly observing runs. An astronomer would look through a microscope eyepiece to monitor a star image forming on a photographic plate located at prime focus. If the image began to blur because of atmospheric refraction or slight flexings in the tube, the scientist would quickly adjust the telescope to get the target object in focus again. It often took tedious hours in the nighttime cold to build up an exposure of a dim star.

Today, astronomers seldom need to work in the prime focus cage. The photographic plate, which had to be changed by hand after each exposure, has been replaced by computer-controlled silicon sensor chips known as charge-coupled devices, or CCDs *(pages 92-93)*. One hundred times more sensitive to light than the most sensitive photographic film, CCDs simply relay their light readings to a computer and are ready for the next target within a few seconds of completing an exposure. Furthermore, the telescope itself can be manipulated from a warm, well-lighted room on the ground floor of the observatory. With CCDs, the Hale's already great sensitivity has improved to the point that it can, in effect, detect a single candle at a distance of 60,000 miles.

One of the Hale's array of CCD cameras, known as Four-Shooter, splits the beam of light, sending it to a quartet of CCDs at the Cassegrain focus. The result is simultaneous images of four adjacent patches of sky. The amount of time it takes to photograph a given area of sky is reduced by this arrangement to a quarter of what it would take with one detector. Similar advances in spectrographic instruments enable the Hale to record information about the movement and chemical composition of stars at twice the pace of older tools and to determine the recessional speed of a hundred galaxies in a cluster simultaneously rather than one at a time.

1: prime focus cage. **2:** prime focus. **3:** prime focus elevator. **4:** telescope tube. **5:** primary mirror. **6:** coudé mirror.

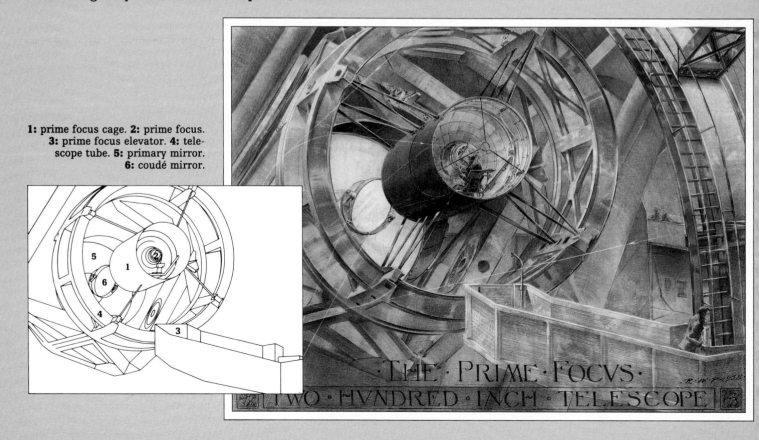

· THE · PRIME · FOCVS ·
TWO · HVNDRED · INCH · TELESCOPE

THE · PRIME FOCUS·
· HOUSING · AND·
· PEDESTAL ·

R.W. PORTER · 1940·

·200 · INCH · TELESCOPE·

Star-forming lanes of dust and hydrogen gas define the spiral arms of the galaxy known as NGC 2997. This false-color image was captured by an electronic light detector called a CCD—or charge- coupled device. The locations of bright young stars are marked in red. The green indicates areas of cooler gas and dust in the spiral arms and older stars in the center of the galaxy.

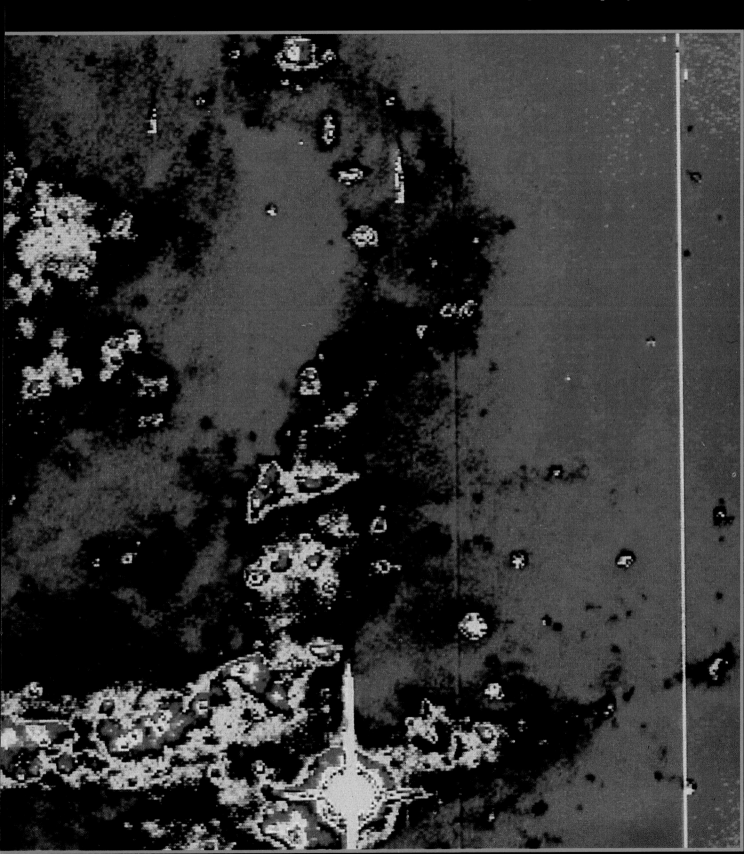

tarlight streaming into the telescopes of the world's major observatories on any clear night is likely to be focused on a sensor called a charge-coupled device. The CCD, as it is more commonly known, is made mostly of silicon. When photons of light strike the silicon atoms, a remarkable thing happens: Light is converted into electrical energy in the form of electrons freed from the silicon atoms. The photons striking the CCD thus generate an electrical charge that is directly proportional to the intensity of the incoming light. The charges are electronically processed to produce images *(pages 90-99)*. Although this photon-to-electron alchemy is only about 80 percent efficient, the silicon component of a CCD is seventy times more sensitive to light than the silver compounds employed in the highest-quality photographic plates, the traditional means of collecting images of the universe.

An invention of the computer age, the CCD caps centuries of progress in techniques for recording celestial observations, beginning with pen and ink. For 200 years and more after the invention of the telescope, astronomers relied on manual notation to log mountains of data in meticulously kept journals. Not only was this a laborious task, but also the results were necessarily incomplete. No astronomer could ever hope to itemize the size and brightness of every object seen through a telescope or note all the spatial relationships between them.

Then came photography. Almost from its birth in the mid-nineteenth century, the new imaging technique promised to be an astronomical boon. As emulsions were improved, photography provided a way to record far more information than could be captured by transcriptions from retina to paper. And because photographic emulsions summed the light striking them, building up an image over the duration of the exposure, they had the potential to reveal celestial features unseeable by the eye.

The attachment of light-sensitive recording devices to telescopes can be traced to a French physicist and sometime scenery painter for the opera, Louis Daguerre. In 1829, attracted by the limited success of other experimenters working with light-sensitive chemicals, he began exploring the imagemaking properties of a copper plate coated with a silver-based emulsion of his own concoction. He first exposed the plate to light through a pinhole pierced in one end of an otherwise light-tight box. Then he bathed the plate in mercury vapor as a developer. Finally, he stabilized the image—which tended to fade—with an application of iodine crystals. In 1839, Monsieur Daguerre announced that his experiments had produced lasting likenesses of objects.

Within the year, Daguerre and Dominique François Arago, director of the Paris Observatory, became the first to use a telescope to make a daguerreotype of a heavenly body—a fuzzy image of the Moon showing less detail than could be seen with the naked eye. During the ensuing decade there followed additional blurry telescopic portraits of the Sun and its surrounding corona of luminous gases, and of the star Vega. But lack of detail was only one shortcoming of the process. Daguerreotypes of objects in space (excepting the Sun) also required long exposures—twenty minutes, for example, to photograph the Moon and much longer for the brightest star.

Soon, astronomers abandoned daguerreotypy for an improved technique called the wet collodion process, which dramatically shortened exposure times. A piece of glass was spread with a solution containing potassium iodide. When the glass was dipped into a bath of silver nitrate, the silver and potassium compounds reacted to form silver iodide, a light-sensitive silver salt. The plate was then exposed while still moist. In 1857, using wet plates in a camera attached to the 15-inch refracting telescope at Harvard College Observatory, astronomer George Bond took a photograph of Vega. Although the image was far from sharp, Bond became an ardent promoter of photography and eloquently propounded its scientific virtues over mere eyesight. "The plates, once secured, can be laid by for future study by daylight and at leisure," he wrote. "The record is there, with no room for doubt or mistake as to its fidelity."

In the 1870s, dry emulsions consisting of silver-salt crystals suspended in hardened gelatin supplanted the wet collodion process. More convenient to work with, dry plates could, for example, be prepared days in advance, whereas wet plates had to be used immediately upon coating. A gelatin plate was also more uniformly responsive to light over its entire surface because the dry emulsion did not gravitate to the lower edge of a plate as the wet variety did. Moreover, this responsiveness remained consistent for the duration of an exposure; the sensitivity of the wet coating, on the other hand, diminished markedly as the collodion mixture dried.

SEEING THE UNSEEABLE

Up to this time, nothing had been captured in an astrophotograph that had not already been observed. But in 1883, British astronomer Ainslee Common yoked a camera to a 36-inch reflecting telescope of his own design and turned the instrument toward the Orion nebula. The resulting picture revealed stars within the gas cloud that had never been seen. A few years later, the French astronomers Paul and Prosper Henry settled a longstanding debate in the scientific community over whether the constellation Pleiades encompassed a nebula. Photographing through a 33-centimeter (13-inch) refractor, they pro-

duced an image that showed wispy clouds enfolding one star of the cluster Merope. Around the same time, the Henrys realized that photographs might also make it easier to identify objects such as asteroids, which move across the sky quickly, relative to stars. In a photograph, they would appear as short streaks rather than dots like the stars and thus could be readily picked out.

The Henrys' successes prodded Edward Barnard, a self-taught astronomer at the Lick Observatory in California, to launch in 1887 one of the first comprehensive photographic mappings of the sky. Over a span of nearly forty years, Barnard produced hundreds of plates, some requiring exposures as long as seven hours. Fifty of the images were published in 1927 as *A Photographic Atlas of Selected Regions of the Milky Way.*

Refractors like the ones that Barnard used for his life's work were well suited to sky surveys, because images of stars observed through them were almost uniformly focused over the telescope's entire field of view—if imperfectly so, owing to aberrations in the lens. However, by the time that Barnard's atlas appeared, large reflecting telescopes, such as the 60-inch and 100-inch instruments at Mount Wilson, were operating. Built with parabolic mirrors that banished the plague of aberrations, these telescopes produced much sharper images than refractors—but only near the center of the mirror, whose curvature blurs objects that lie more than a fraction of an inch from its optical axis, a line perpendicular to the center of the mirror. The sharply focused area of the Hale telescope, for example, is only 60 arc seconds in diameter, or about the size of a baby lima bean. For a view of the heavens comparable to those produced by telescopes with lenses, astronomers using reflectors had to laboriously assemble a mosaic made from the centers of thousands of images.

Salvation from this tedium came at the hands of an eccentric Norwegian lens crafter named Bernhard Schmidt. Born in 1879, Schmidt spent his early years on an island off the coast of Estonia called Naissaar (Island of Women). The five-mile-long sliver of land in the Baltic Sea, which had appeared on some medieval maps as Amazonia or Terra Feminarum, was home to fewer than 500 people, mostly fishermen, plus a few farmers who tilled the coastal perimeter of a swampy interior. Here, Bernhard led a quiet childhood—with

Dust and gases of the Orion nebula appear as a shapeless fog punctuated by brilliant stars in this hour-long exposure, made in 1880 at the Harvard College Observatory by Henry Draper *(inset).* Among the instruments Draper developed was a multi-prism spectrograph that allowed an astronomer to make as many as nine spectrograms at a time.

one momentous exception that stemmed from his fascination with rocketry. On a Sunday morning in his fifteenth year, when his parents were in church, he launched a homemade missile consisting of a metal pipe filled with gunpowder. Instead of rising like a rocket, the device exploded like a bomb, taking off Bernhard's right hand and part of his arm.

Thereafter, Schmidt took up the grinding of telescope lenses as a hobby. Following his graduation from secondary school in Sweden, he moved to Mittweida, Germany, where he spent a short period of time as a student at the local technical institute. During his early twenties, he set himself up as an independent lens grinder in an abandoned bowling alley behind Bretschneider's Lindengarten tavern and restaurant. On clear nights, Schmidt would shuttle back and forth between his shop and a telescope he had put up in a vacant lot across the street from the restaurant, pouring himself a brandy on each passing of the bar. "We used to make out very well with him," Frau Bretschneider reported.

In his shop, dressed formally in striped trousers and a cutaway coat with the right sleeve pinned up, Schmidt spent long nights polishing disks of glass to form telescope mirrors that he sold to astronomers throughout Europe. His specialty was the paraboloid mirror. Although there were machines for such work, Schmidt insisted that they produced flawed results. His hand, he claimed, was "more sensitive than any gauge." In the years leading up to World War I, his expertise earned him a reputation in the astronomical community as a genius.

A pacifist before the war, Schmidt became a misanthrope, convinced by the conflict that men acting in any number greater than one presented a danger: Two would quarrel, a hundred would make a rabble, and a thousand would start a war. Already shy, he became more withdrawn. After the war, he moved to a small town outside Hamburg at the invitation of the Hamburg Observatory. There, he lived in a men's dormitory and figured lenses for observatory scientists, who referred to him respectfully as *der Optiker*—the optician.

A REVOLUTIONARY CAMERA

In 1929, Schmidt and a young astronomer, Walter Baade, made an excursion at the behest of the observatory to photograph a solar eclipse from a vantage point in the Philippines. During the long journey by sea, Schmidt mentioned to Baade his idea for a combination reflecting telescope and camera that would take sharp pictures covering broad expanses of sky. His concept had two parts: First he would grind a special lens to be placed in front of the telescope mirror; the lens would slightly bend the incoming light to help correct the mirror's tendency to blur objects that lay too far from the optical axis. Schmidt opted for a spherical mirror because it was a simpler shape to grind than a paraboloid and because spherical aberration was easier to correct with his lens than the severe coma aberration characteristic of parabolic reflectors. To finish the job, he would make a holder for photographic film or plates that would bow the glass or film slightly toward the mirror, giving the plate a subtle convex curve. With these improvements, Schmidt hoped to capture images spanning as many as 10 degrees of arc but having no more fuzziness than those produced at the center of the best mirrors.

Intoxicated by the possibilities, Baade urged Schmidt to build a prototype without delay, but Schmidt demurred, saying that he had first to devise an elegant technique for grinding the unusually shaped correcting lens, which was to be about two-thirds the diameter of the telescope's mirror. Lens curvature would be mostly spherical. But unlike most lenses, which are thinnest around the perimeter, Schmidt's corrective glass would thicken slightly at the edge in order to perform as he intended. In less than a year, he came up with a suitable polishing technique, and by 1931 the first Schmidt camera was installed at Hamburg. Immediately, it produced photographs filled with stunning shoals of sharply focused stars.

HERE A SCHMIDT, THERE A SCHMIDT

Der Optiker died in 1935. The following year, Hamburg Observatory made public the configuration of his remarkable invention, enabling astronomers around the world to replicate the device. In 1936, a Schmidt camera, nick-named Little Schmidt, was installed atop Palomar Mountain in California. Built at the suggestion of Walter Baade, who several years earlier had immigrated from Germany, Little Schmidt had an 18-inch correcting lens in front of a 26-inch mirror. A Caltech astronomer named Fritz Zwicky would soon put it to work searching for exploding stars that he and Baade called supernovae.

In 1933, the two astronomers had proposed the name for a seemingly rare, exceptionally bright, and long-lasting variety of stellar explosion. Other astronomers usually grouped them with novae, much less brilliant and relatively short-lived events experienced, sometimes repeatedly, by many stars the size of the Sun as they begin to flicker out. Zwicky and Baade, however, considered supernovae to be a different kind of phenomenon—the final cataclysms of much larger stars.

Few astronomers supported their view, arguing that the so-called super-novae were too uncommon to represent the deaths of an entire class of stars. So, to prove that supernovae were not so infrequent as widely believed, Zwicky began a quest for examples of the phenomenon in a cluster of galaxies near the constellation Virgo. Empty-handed after a two-year effort with a small telescope, he persuaded the director of Palomar Observatory to let him use Little Schmidt to search broader expanses of the heavens. In March 1937, within months of renewing the pursuit, the astronomer recorded a bright burst in the spiral galaxy NGC 4157 that fit his profile of a supernova. Other sightings soon followed.

A decade after Little Schmidt went into service, it was joined by Big Schmidt. This instrument had a 72-inch mirror and a 48-inch correcting lens, larger than the 40-inch maximum established for refractor objectives by the Yerkes telescope. The Schmidt lens could exceed this limit only because, being a corrective lens, it was considerably thinner and weighed substantially less.

Beginning in 1949, a team of researchers backed by Caltech and the National Geographic Society mustered the newer of the two Schmidt cameras at Palomar for an ambitious undertaking—to survey the northern sky. This project, which took five years to complete and remains unrivaled for comprehensiveness, filled 1,870 plates and revealed stars a million times dimmer than can be seen by the naked eye.

Notwithstanding virtuoso performances of Big Schmidt and other Schmidt cameras, photography had reached a plateau by the late 1950s. Though photographic emulsions were capable of translating many thousands of brightness levels into corresponding tones, the relation between brilliance in the source and tones of gray in a negative was nonlinear. That is, one star, twice as bright as another, did not produce twice the density in the negative—except over the narrow middle of a plate's tonal range. This unavoidable characteristic of photographic plates tended to reduce astronomers' confidence in their ability to judge the full range of brightness represented in a plate.

French astronomers Paul and Prosper Henry ready their astrographic camera for an exposure. Built in 1885 at the Paris Observatory, the camera consisted of two lenses placed side by side within the telescope. One lens was used to aim the camera; the other exposed a photographic plate.

Furthermore, emulsion speed—a measure of its sensitivity to light—had improved only marginally since the turn of the century. The same was true of resolution, the ability of a plate to record fine detail. Even the most sensitive plates still generally registered less than one percent of incoming photons. A plate might be exposed longer in an effort to gather more light, but only at the cost of emphasizing the masking effects of airglow, nighttime emanations from the upper atmosphere as it releases energy absorbed from sunlight during the day.

Slow emulsions were all the more disappointing in the field of spectroscopy, where much activity in astronomy had become centered since the 1930s. To create a spectrum for examination, light from a star is spread out to reveal its constituent colors. Doing so reduces the intensity of the light falling at a given point on a plate, limiting the ability of astronomers to study dim stars spectrographically. The dimmest stars suitable for spectroscopy with the 200-inch Hale Telescope, for example, were nearly sixteen times brighter than the faintest objects photographable with the instrument.

A SURGE OF PROGRESS

Little progress was made until the 1970s, when significant advances came on two separate fronts—photographic chemistry and computers. In photography, a battery of new emulsions appeared on the market. Some were concocted to respond most strongly to green wavelengths, so that they would register faint stellar objects but not the red wavelengths that predominate in airglow. Others offered improved resolution from finer grains of silver salts. These emulsions in particular exacted a price in reduced sensitivity, because the smaller grains are less responsive to light than larger ones. To compensate, astronomers formulated ways to hypersensitize these emulsions, typically boosting their responsiveness by about five times, and occasionally as much as 2,000 times. The exotic procedures involved baking plates while exposing them to gases such as nitrogen or hydrogen, or to the vapors of ammonia or mercury. The object was to drive from the emulsion's gelatin any oxygen or moisture, either of which lowered the plate's responsiveness to light. At Palomar Observatory, a special building dedicated to ''hypering'' emulsions was jokingly dubbed the Hindenburg Cottage, in reference to the flammable nature of the hydrogen gas used in the process.

Even as photographic emulsions were improving, darkroom expert David Malin, head of the photo lab at the Anglo-Australian Observatory in New South Wales, applied every trick he knew in his efforts to counteract the effects of airglow and to otherwise manipulate astrophotographic plates to extract the most information possible. One difficulty with photographic plates, for example, is that they print poorly on photographic paper. In the developed plate—a negative transparency in which light areas appear dark—the brightest areas might be nearly 10,000 times denser than the dimmest areas. Photographic papers, however, commonly register a range in brightness of perhaps 50 to 1. Depending on

exposure time, a print of an astrophotographic negative on such paper could be made to reveal bright details, dim ones, or some in the middle, but not the full range simultaneously.

To solve the problem, Malin adopted a technique invented by lithographers during the 1930s. Called unsharp masking, the three-step darkroom procedure yields, almost magically, a new negative in which the brightest details are only about thirty times more brilliant than the dimmest, a tonal range that photographic papers accommodate with ease.

Using this approach, Malin extracted from old plates information that had not been previously accessible. In one case, he applied his expertise to photographs taken with the United Kingdom's Schmidt telescope in Siding Springs, Australia, and made out a galaxy, theretofore unsuspected, lying in the direction of the constellation Virgo. Spectrographic examination of the galaxy, given the name Malin 1, revealed that it is nearly one billion light-years from Earth. Having established the distance, astronomers then made measurements that ranked the galaxy as one of the most massive and luminous star systems ever detected.

The enormous amount of information stored on photographs of the heavens (a single image taken by a Schmidt camera might encompass as many as a million stars) has always been almost too much of a good thing. Having the data was a comfort, but there were so many stars that searching manually through a sky survey to catalog examples of a particular kind of star or galaxy, for instance, was a numbingly monotonous task, poorly suited to an astronomer's brain. Some kind of machine was needed, and that machine was the computer.

The key to the use of computers for this purpose is a process called digitizing, in which the information in the image is converted into strings of numbers that a computer can easily store for later retrieval and analysis. Digitizing a plate involves dividing it into a fine-meshed grid made up of thousands of squares called picture elements, or pixels. A beam of light is passed through each pixel of the image, one at a time, and the intensity of the beam is measured as it emerges from the other side. Each grid square is assigned a number, typically consisting of eight ones and zeros, or bits, the lingua franca of digital computers, that denote its brightness on a scale of 0 to 255 and its position in the grid.

In the late 1960s, the crucial element in the process of transferring information from plates to computers was an electronic device called a photomultiplier tube, employed to measure the intensity of the light beam passed through the image on a photographic plate. Widely used in instruments for detecting nuclear radiation and measuring very dim light, for example, the photomultiplier is a type of vacuum tube similar in principle to components found in radios and television sets before transistors became common. Developed in the late 1930s,

Estonian-born Bernhard Schmidt—pictured here in his optics laboratory at the Hamburg Observatory in 1928—gave astronomy a valuable new viewing instrument when he fitted a combination camera and reflecting telescope with a correcting glass plate that allowed sharp images over a wide field of view. The circular device on the right is a guide for the mirror-polishing machine at left.

the photomultiplier tube relies on a phenomenon known as the photoelectric effect, in which photons of light, upon colliding with certain metallic elements and compounds, create an electric current.

The device is, in essence, a light amplifier that consists of a chain of electrodes. At one end is a negatively charged photocathode. In photomultipliers used in astronomy, this electrode is usually a glass plate spread with a mixture of antimony and cesium. Next in succession is a sequence of similarly coated, positively charged electrodes called dynodes. The chain of electrodes ends with a positively charged plate called an anode.

Photons entering the tube strike atoms of the photocathode's antimony-cesium coating. The collisions trigger the ejection of electrons, creating a tiny current that varies with the intensity of the light. Accelerated by the attraction of opposite charges, the electrons crash into the first dynode, where each spawns several offspring. They are in turn speeded up, then multiplied in collisions with the next dynode. After several such stages, the initial burst of electrons is amplified about 100 times.

This current, which is proportional to the transparency of the plate at the point being measured, is assigned a number that is stored in a computer as a brightness value for a pixel. Digitizing a single astrophotograph requires many thousands of such samplings.

Once images had been translated into numbers, they could be manipulated by computer software. During the 1970s, systems such as the Galaxy Machine at Scotland's Royal Observatory and the Automatic Plate Measuring Machine in Cambridge, England, were programmed to compare brightness data about groups of pixels with the known light output of different kinds of celestial objects. Clever mathematical formulas called algorithms—actually instructions for the computer—permitted the machine to distinguish between stars, galaxies, comets, and nebulae, or to search an entire photograph for a single type of object. In 1985, researchers at the National Optical Astronomy Observatories in Arizona developed an image-processing program called IRAF (Image Reduction and Analysis Facility) that could extract information about an object's size and shape. Systems like Galaxy Machine and IRAF, which sprang from astronomical laboratories around the world, revitalized photography as a useful tool in a way that few could have imagined less than a quarter-century earlier.

SUBSTITUTING FOR PLATES

The photomultiplier tube's contribution to astronomy would no doubt be considered noteworthy even if it had been used only for digitization, but it also had three significant advantages over photographic emulsions. First, over a wide range of tones, a photomultiplier tube generated currents in direct proportion to the number of photons falling on the photocathode. Second, by amplifying light, the new device could detect faint objects in space that would be missed by a conventional plate, thus making possible the discovery of new stars and the spectroscopic analysis of familiar ones having

spectra too dim to capture photographically. Finally, a photomultiplier tube could register in a single exposure objects up to 100,000 times fainter than the brightest ones in the picture. A photographic negative could handle a brightness range of only 10,000 to 1.

AN UNEXPLOITED VERSATILITY

In 1969, the sensitivity of another type of vacuum-tube light detector, called an image intensifier, was proposed for a different role—to make short exposures of bright objects rather than long exposures of faint ones. The author of this notion was Antoine Labeyrie, a French physicist in the field of optics at the State University of New York at Stony Brook, Long Island. For some time, Labeyrie had been contemplating the phenomenon of twinkling caused by the atmosphere. Twinkling is the bane of astronomers. It results from thermal variations in the atmosphere that cause stars to change their apparent brightness and sometimes to shift position *(pages 76-77)*. The effect is to reduce the resolution of even the best telescopes to a fraction of what they could achieve in the absence of an atmosphere.

Labeyrie suggested that light from a star passing through this shell of gases, smoke, and dust undergoes a transformation best approximated by an everyday experience at a swimming pool. An object on the bottom—a penny, say—appears to be a flickering of many pennies when seen through turbulent water, defying any attempt to fix on the real one. Just as a short-exposure photograph of the pool surface would freeze this flickering, Labeyrie surmised that a similar picture of a star would capture a twinkling star as flecks of light. He called them speckles. By combining speckles mathematically in a computer, he reasoned, it would be possible to reassemble a single image having an unprecedented degree of resolution. Labeyrie calculated that, to accomplish his goal, he would need an exposure of one-thirtieth of a second or less. Such a feat was well beyond the capabilities of the most sensitive plate, even with

A Schmidt telescope and a darkroom technique called unsharp masking produced this image of the Orion nebula. First used in the mid-1970s by Australian astronomer David Malin *(right),* unsharp masking combines a high-contrast negative with a lower-contrast positive to produce an image with a narrower contrast range, thus revealing faint details that would otherwise be lost.

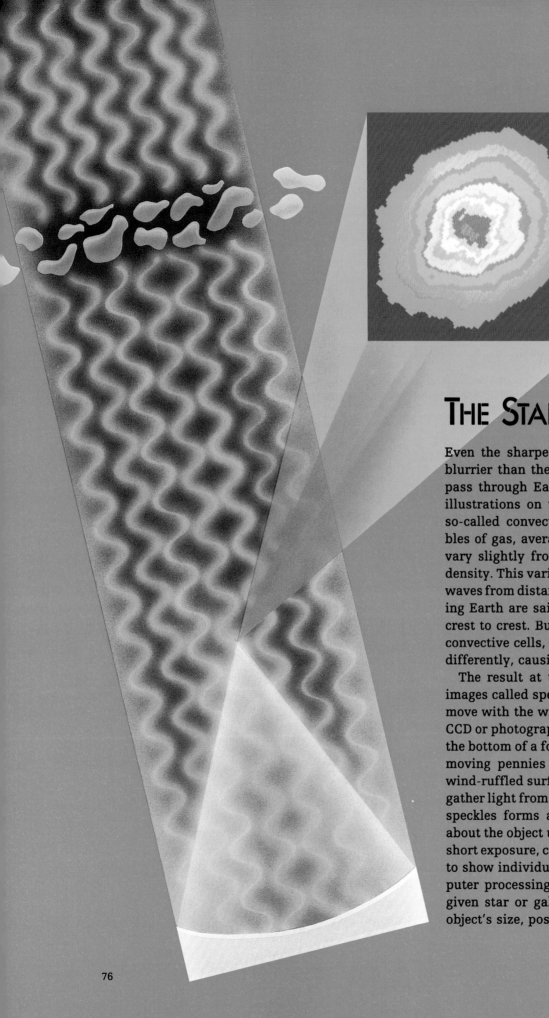

In this ten-second exposure of the binary star Capella, any indication of the stellar pair is obscured by atmospheric blurring. The image, made with a CCD camera, has been colored by computer to reveal increasing levels of brightness.

THE STARS UNBLURRED

Even the sharpest photographs of the heavens are blurrier than they would be if light did not have to pass through Earth's atmosphere. As shown in the illustrations on these pages, the air is made up of so-called convective cells—irregularly shaped bubbles of gas, averaging just a few inches across, that vary slightly from one another in temperature and density. This variation has a disruptive effect on light waves from distant celestial objects. Waves approaching Earth are said to be "in phase," that is, aligned crest to crest. But as they pass through a column of convective cells, the cells each refract, or bend, light differently, causing the waves to slip out of phase.

The result at the telescope is multiple, distorted images called speckles. Moreover, as convective cells move with the winds aloft, speckles shift about on a CCD or photographic plate, much as a single penny at the bottom of a fountain appears to be multiple, ever-moving pennies when viewed through the water's wind-ruffled surface. In the long exposures needed to gather light from greatly distant sources, the crowd of speckles forms an amorphous blob that tells little about the object under scrutiny *(above)*. In contrast, a short exposure, called a specklegram, stops the action to show individual speckles *(right)*. Subsequent computer processing of hundreds of specklegrams of a given star or galaxy can reveal such details as the object's size, position, and structure.

Made with an exposure lasting one-sixtieth of a second, this specklegram of Capella *(below)* reveals pairs of speckles *(right)* that are created when convective cells in the atmosphere distort and multiply the binary star's image. This photograph is atypical, however. More often, the large number and random positioning of the individual speckles act to conceal any pairing or other relationship between them.

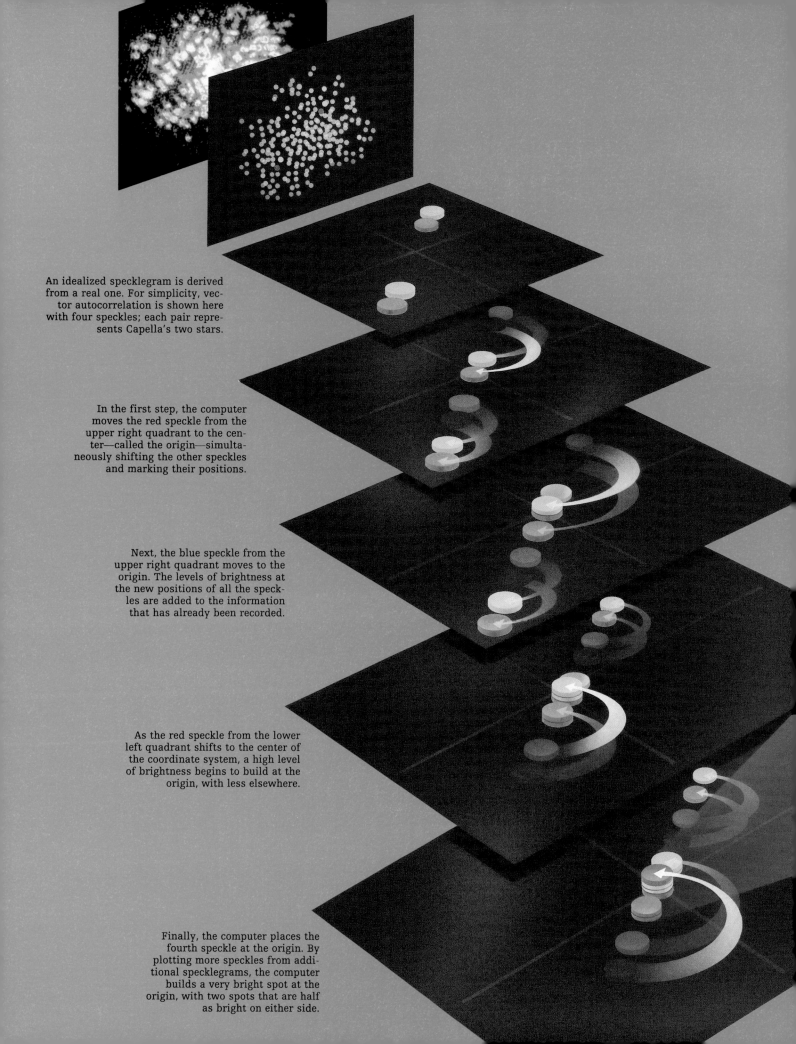

An idealized specklegram is derived from a real one. For simplicity, vector autocorrelation is shown here with four speckles; each pair represents Capella's two stars.

In the first step, the computer moves the red speckle from the upper right quadrant to the center—called the origin—simultaneously shifting the other speckles and marking their positions.

Next, the blue speckle from the upper right quadrant moves to the origin. The levels of brightness at the new positions of all the speckles are added to the information that has already been recorded.

As the red speckle from the lower left quadrant shifts to the center of the coordinate system, a high level of brightness begins to build at the origin, with less elsewhere.

Finally, the computer places the fourth speckle at the origin. By plotting more speckles from additional specklegrams, the computer builds a very bright spot at the origin, with two spots that are half as bright on either side.

Unraveling Clouds of Speckles

Using a mathematical technique called vector autocorrelation, astronomers are able to create clear representations of celestial objects. A high-resolution composite image can be structured from the raw material of hundreds or even thousands of specklegrams, which are processed one at a time. First, brightness levels from a specklegram are loaded into a computer's memory. The machine then constructs a kind of idealized version of the specklegram by recording the coordinates of only those pixels that exceed a certain brightness threshold. The pixels are deemed likely locations within each speckle of the object that is being investigated.

Next, the computer repeatedly shifts the entire pattern of speckles so that each lands in turn at the origin, or center, of a coordinate system. After every move, the computer measures and records the brightness of every pixel.

By keeping a running total of the light intensities at each location in the system, the computer gradually builds a map that shows the object's discrete structure superimposed on a background that represents the effects of atmospheric blurring. (For clarity, the background is not shown in the simplified illustration of the process on the opposite page.)

The result of the process is called a vector autocorrelogram. For the binary star Capella *(below)*, it shows a bright spot flanked by two dimmer ones. With three elements instead of two, the vector autocorrelogram is clearly not a true image of Capella. It shows only the orientation of the binary system's members to each other and their angular separation. Determining which spot represents which star—and which is an artifact of the process—requires additional analysis that can include reconstructing real images, as shown on the following pages.

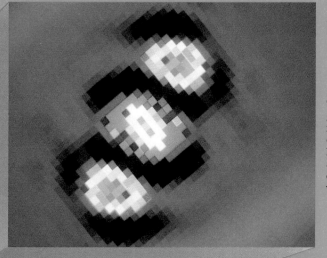

The completed vector autocorrelogram, containing position data from many specklegrams, is falsely colored according to brightness. Because of the way that speckles are shifted during the process, one of the flanking images is the inverted mirror image of the other.

The raw material for speckle imaging is digitized information taken from many specklegrams *(left)*. Processing the data from each specklegram in turn, a computer uses Fourier analysis to obtain brightness maps *(below, top)* and applies the Knox-Thompson algorithm to construct phase maps *(below, bottom)*.

In the second stage of the imaging process, the computer integrates phase and brightness maps to produce a single high-resolution image that correctly renders both the shapes and the positions of objects. The image of a triple star at right, taken with the 2.3-meter (90-inch) reflector at Steward Observatory on Kitt Peak in Arizona, resolves details separated by less than .05 arc second—nearly the resolution that would be possible with a same-size telescope from a vantage point above the atmosphere.

PICTURES OF SURPASSING CLARITY

Although each speckle in a specklegram is a small, complete image, enlarging only magnifies the distortion caused by convective cells in the atmosphere, without providing any new information. With a technique called speckle imaging, however, the speckles used for vector autocorrelation can be made to yield images with resolution as high as one sixty-sixth of an arc second, fifteen times finer than is possible with ordinary long-exposure imaging.

Undoing the corruption of light waves that results when the waves pass through the atmosphere is a process even more computer intensive than vector autocorrelation. In addition to the brightness and position information recorded in the specklegram, information about the phase of the light in each of the speckles is also needed. Since a CCD image records no such data, phase must be estimated. One common method for doing so is called the Knox-Thompson algorithm. (An algorithm is a kind of recipe for a computer to follow toward a solution to a problem.) Named for its creators, Keith Knox of the Xerox Corpora-

tion in Webster, New York, and Brian J. Thompson, provost of the University of Rochester, the algorithm uses a mathematical technique called Fourier analysis to make the necessary phase approximations. In essence, Fourier analysis is a process by which complex waveforms—such as those that combine to make the distinctive sound of a musical instrument or to create the distortion seen in speckles—can be disassembled into their simpler components. The reverse process—from components to complex wave—is also possible.

Programmed with this algorithm, the computer works backward from specklegrams to gauge how light must have been distorted, while passing through the atmosphere, in order for the speckles to occupy the positions they do in the specklegram. For each of many specklegrams, the algorithm produces an array of numbers called a phase map. Combined with brightness maps—complementary numerical arrays that are also produced by Fourier analysis of speckle data—the phase maps can produce otherwise unobtainable, high-resolution views of celestial phenomena ranging from the triple star shown here to asteroids and the vast remnants of exploded stars pictured overleaf.

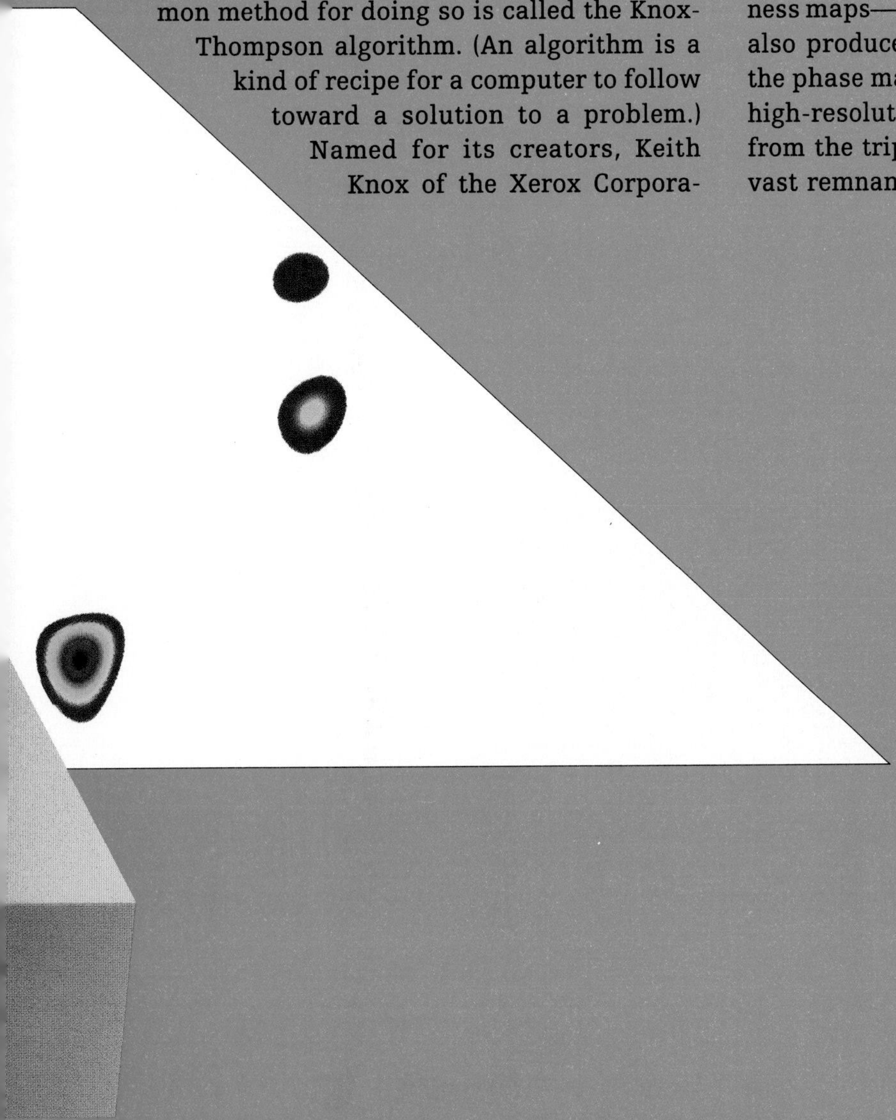

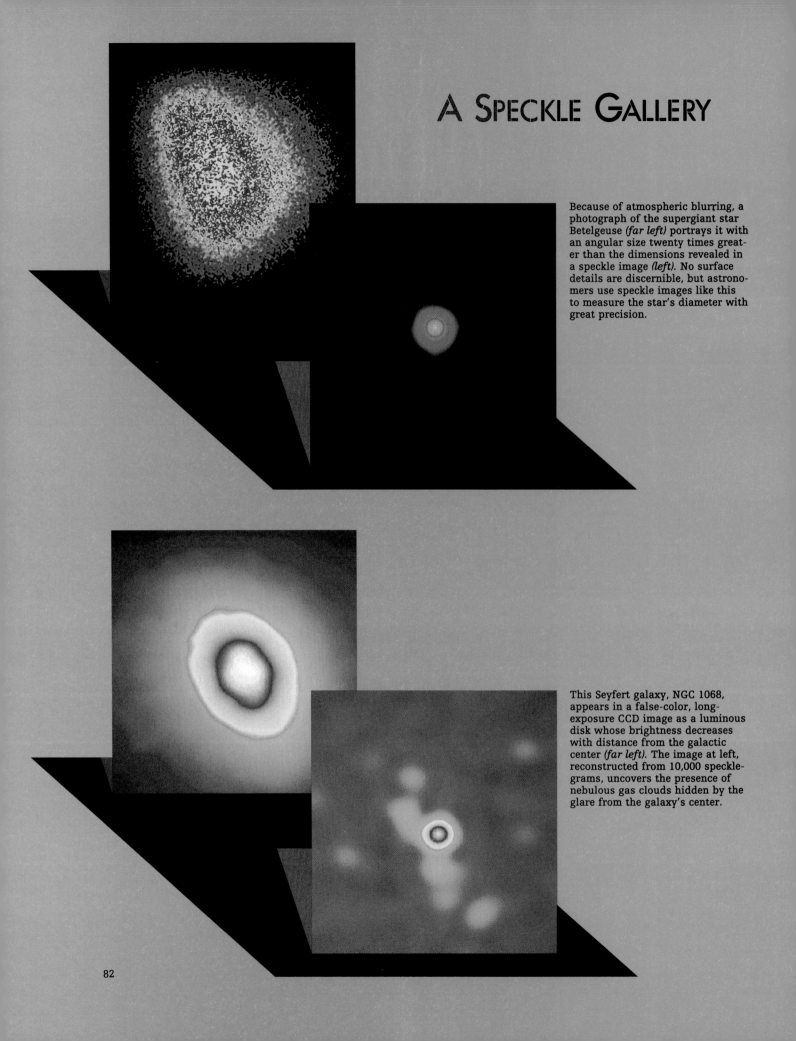

A Speckle Gallery

Because of atmospheric blurring, a photograph of the supergiant star Betelgeuse *(far left)* portrays it with an angular size twenty times greater than the dimensions revealed in a speckle image *(left)*. No surface details are discernible, but astronomers use speckle images like this to measure the star's diameter with great precision.

This Seyfert galaxy, NGC 1068, appears in a false-color, long-exposure CCD image as a luminous disk whose brightness decreases with distance from the galactic center *(far left)*. The image at left, reconstructed from 10,000 speckle-grams, uncovers the presence of nebulous gas clouds hidden by the glare from the galaxy's center.

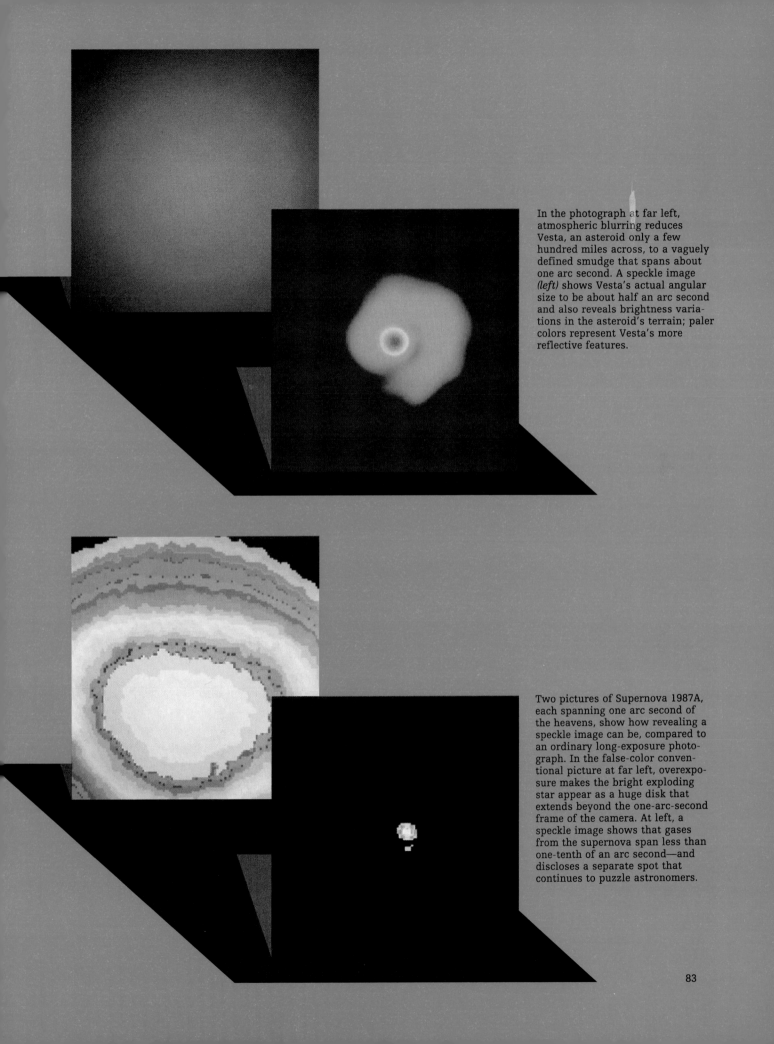

In the photograph at far left, atmospheric blurring reduces Vesta, an asteroid only a few hundred miles across, to a vaguely defined smudge that spans about one arc second. A speckle image *(left)* shows Vesta's actual angular size to be about half an arc second and also reveals brightness variations in the asteroid's terrain; paler colors represent Vesta's more reflective features.

Two pictures of Supernova 1987A, each spanning one arc second of the heavens, show how revealing a speckle image can be, compared to an ordinary long-exposure photograph. In the false-color conventional picture at far left, overexposure makes the bright exploding star appear as a huge disk that extends beyond the one-arc-second frame of the camera. At left, a speckle image shows that gases from the supernova span less than one-tenth of an arc second—and discloses a separate spot that continues to puzzle astronomers.

the telescope aimed at very bright stars. So Labeyrie proposed assembling a battery of image intensifiers to brighten the speckles sufficiently to photograph with a specially modified movie camera. Operating at normal speed, such a camera has a shutter speed of about one twenty-fourth of a second.

Seeking to test his theory, Labeyrie wrote to Horace Babcock, then director of the Palomar Observatory, suggesting that someone there attempt to observe speckles using the Hale Telescope. Babcock contacted the astronomy department at Stony Brook to check the credentials of this "crazy French guy," whom no astronomer seemed to have heard of. Professors in the department dispatched a delegation—Professor Stephen Strom and two graduate students, Robert Stachnik and Daniel Gezari—to meet with the physicist.

After interviewing Labeyrie, the astronomers were persuaded that his startlingly original idea had real merit. Stachnik and Gezari prepared a proposal for an experiment and won time for it on the Hale—odd hours at dawn and dusk that were in little demand by other astronomers. Sure enough, over several anxious sessions atop Palomar Mountain in 1970, they succeeded in recording speckle patterns, just as Labeyrie had predicted. Computer programs for consolidating the data soon followed.

With atmospheric blurring largely eliminated, astronomers could glean critical information about stars' true diameters, positions, and motions. From such data, the mass of a star could be derived. During their stint on the Hale, Stachnik and Gezari were able to measure the width of several stars, and soon other astronomers who adopted the technique developed additional computer software that processed speckles into striking new images of not only stars but also the Sun, asteroids, and the outer planets and their satellites *(pages 78-83)*.

A MEMORY FOR LIGHT

The year before Stachnik and Gezari confirmed Labeyrie's speckle theory, computer researchers Willard Boyle and George Smith at Bell Telephone Laboratories had begun to look for a way to pack more data onto memory chips, which store the electrical charges that encode digital information as ones and zeros. The solution that the two inventors hit upon was the charge-coupled device. Smith later recalled that he and Boyle "started batting ideas around and invented it so quickly that we surprised ourselves." Within a decade, their invention would have a profound impact on astronomy.

Like other chips in computers, the CCD was to be made of silicon. But instead of giving the chip a separate, microscopic circuit for each of thousands of individual memory cells that would store a charge, Smith and Boyle came up with a design that would require little wiring: The charge would be contained in tiny structures that the two inventors called potential wells. In essence, a potential well was a holding area for electrons, electrically isolated from adjacent wells. By shifting the barriers across the chip electronically through simple circuitry along its edges, each charge could be moved off the chip and into a computer's processing area—somewhat as if the electrons were carried along like marbles trapped between a pair of rollers moving across a table. Within a few weeks, the researchers had a prototype in hand. It worked just as they had expected.

By nature, a CCD yields its cache of data only serially. That is, the computer can retrieve a distant bit only by waiting for all the intervening ones to pass in review, a time-consuming process by computer-speed standards. The circuitry on a conventional chip, by contrast, permits virtually instant access to any bit stored there, regardless of the memory cell it occupies. Because computer designers usually opt for speed over compactness, the CCD languished as a memory chip. On another front, however, where the relatively slow speed of serial access was acceptable, the CCD proved to be a virtually unqualified success.

SELLING THE CCD TO ASTRONOMERS

Among potential applications for CCDs listed by Boyle and Smith in their paper describing the invention was its use as a detector of light. Silicon, the base material of a CCD, is one of those few substances that exhibit the photoelectric effect. The two inventors thus recognized that photons of light falling on the chip could be transformed into electrons occupying a potential well. Like the photomultiplier tube, the CCD could record a large brightness range; present-day CCDs have a range of 100,000 to 1. The chip also emitted electrons that increased in lockstep with the number of photons striking it and thus displayed not the slightest trace of nonlinearity. But it required less power to operate than a photomultiplier tube, and it was ten times as sensitive to light.

Quick to exploit the CCD as a means of making electronic images, Bell Laboratories in 1971 unveiled a prototype black-and-white TV camera built around a charge-coupled device. A color camera followed a year later.

Around the same time, scientists at NASA's Jet Propulsion Laboratory in Pasadena, having heard about the CCD, envisioned a berth for the sensor on planetary space probes. After some initial study by JPL engineer Fred Landauer, who had worked with imaging systems aboard NASA's Ranger and Surveyor lunar probes, JPL hired James Janesick, a former U.S. Navy electronics technician—and one-time lead guitarist for a rock band called the Tangents. Charge-coupled devices were so new that few scientists or engineers knew much about them, and Janesick was not among the initiates. But in no time he brought himself up to speed.

Janesick became convinced that CCDs represented the wave of the future, likely to be the most important astronomical advance since the photograph. Here was an almost perfect detector. Sensitive to wavelengths from the near infrared to x-rays, it could tally photons with ease, rarely ringing up spurious signals. Each of its several thousand pixels snagged almost every photon falling upon it. In a single exposure, it recorded the emanations of the brightest celestial objects as well as those 10,000 times fainter, allowing astronomers to peer ten times deeper into space than before. The CCD's only bothersome aspect was its size; not much larger than a thumbnail, it could make images of only a small portion of the sky with each exposure.

Hoping to win places for CCDs on several upcoming NASA missions, Janesick in 1966 persuaded Caltech engineer James Westphal to help him build a camera fitted with the device. Janesick's intention was to demonstrate the instrument around the observatory circuit to show what a CCD could do when attached to a telescope. "Marketing," the pragmatic Janesick called the enterprise. Without such a selling job, he felt, much time would pass before the conservative community of astronomers appreciated the CCD's merits.

The camera impressed its audience, but not until 1975 did Janesick cement the CCD's future. He and a team of scientists from JPL and the University of Arizona hooked a second-generation CCD camera to the 61-inch reflector atop Mount Lemmon outside Tucson to demonstrate the superiority of the CCD's powers of resolution compared with a photographic plate. Focusing on Uranus, Janesick and crew captured the planet as it might appear if photographed through a telescope twice as large as the Hale. For the first time, an accurate diameter of the planet was taken, and the images hinted at mysterious happenings in the atmosphere above the planet's south pole. Eleven years later, the *Voyager 2* probe found these to be high methane clouds. Soon after the Mount Lemmon success, NASA announced a competition to determine the best sort of camera for the 94-inch Hubble Space Telescope. In response, Westphal purchased an eight-dollar spaghetti pot and built a CCD camera inside it. Christened JPL-11 and attached to the Hale Telescope for a demonstration, it defeated two non-CCD competitors.

Largely responsible for bringing CCDs to astrophotography in the mid-1970s, Jet Propulsion Laboratory scientist James Janesick is now researching ways to protect CCDs aboard space-based telescopes from the intense radiation found beyond Earth's atmosphere.

By the mid-1970s, the CCD had become essential to astronomy. No first-rate observatory could be without a CCD camera. Palomar Observatory, for instance, acquired several of them. One, built in 1976, was known formally as the Prime Focus Universal Extragalactic Instrument (PFUEI). Its CCD was among the most sensitive of the day and, with an 800 x 800 array of pixels, among the highest in resolution. Aiming PFUEI toward a quasar in 1979, a team of British and American astronomers used the camera's superior resolution to prove the existence of a so-called gravitational lens whose presence they had first suspected from an indistinct photographic image taken through the 2.3-meter (90-inch) telescope at Arizona's Steward Observatory. Lensing, predicted in the 1930s from Einstein's general theory of relativity, is a rare phenomenon in which light from a distant galaxy, for example, passes through a nearer galaxy's enormous gravitational field, which splits the light into multiple, magnified images of the source.

The PFUEI proved to be most helpful in everyday spectroscopy, especially of very faint stars and galaxies. In such studies, astronomers spread a star's light into a spectrum with a diffraction grating—a glass plate superimposed with a plastic film etched with fine lines a few thousandths of a millimeter apart. To exclude extraneous light from nearby stars, a thin sheet of metal cut with a narrow slit—a tenth of a millimeter across—is placed in front of the diffraction grating. For a spectrum to be created, a telescope must be aimed so that light from a target star passes through the slit. Pointing the telescope at any object so dim as to be invisible required fixing on a set of coordinates in the sky where the object was known to be. Before PFUEI, this was a hit-or-miss affair, because aiming by coordinates was not precise enough to assure that light from the star passed through the diffraction grating slit. Only after an exposure was made would an astronomer know whether a night's work had been wasted. The sensitivity of PFUEI's charge-coupled device, however, permitted an astronomer to see the target on a video monitor while aiming the telescope. The quarry thus was caught every time.

MR. FIX-IT

The Prime Focus Universal Extragalactic Instrument was the creation of James Gunn—theorist, stargazer, and tinkerer extraordinaire. Gunn was born in 1938 and acquired from his father, a peripatetic prospector for Gulf Oil Corporation who died when Jim was eleven, a knowledge of drill presses, lathes, acetylene torches, and the world of tools. At eight, with his father's help, he cobbled together his first telescope, using a mailing tube, a lens from a pair of discarded eyeglasses, and an eyepiece taken from an inexpensive microscope. Gunn's youth was frequently punctuated by a chain of scientific creations and adventures, most of them involving homemade rocket engines.

In one incident, he and a friend named Bill Davis ordered nitric acid and liquid aniline from a mail-order chemical supply house. Mixed together, these toxic chemicals (aniline is a nerve poison that can kill on contact with the skin) catch fire spontaneously, then burn violently. Seeking maximum potency in his fuel, Gunn distilled the nitric acid into a more concentrated form known as red fuming nitric acid. The two friends built a combustion chamber for the fuel and a nozzle for the engine's expected exhaust gases, then anchored the engine to a wooden workbench in a vacant lot behind a sporting goods store owned by Davis's father. After coupling a tank of aniline and a tank of acid to the engine, they opened valves to admit the chemicals to the combustion chamber. The unexpected result was a huge fireball that consumed the entire experiment, workbench and all. Fortunately, the two experimenters escaped uninjured, but shortly thereafter Gunn switched hobbies in much the same way Bernhard Schmidt had half a century earlier: He took up astronomical optics, grinding his first telescope mirror from a Pyrex blank. "Once I got into mirrors," Gunn recalled, "I stopped playing around with explosives."

Both as an undergraduate at Rice University in Houston, where he majored in physics and mathematics, and later as a graduate student at Caltech, Gunn was chiefly interested in exploring the theoretical fine points of cosmology. But Gunn could not keep away from the workbench, and he tackled increasingly ambitious projects. From simple reflecting telescopes he graduated to motor-driven models fitted with assorted cameras, and thence to instruments of entirely original design. His earliest invention, dubbed Gunn's First Machine by fellow students at Caltech, was completed in 1963. It resembled a filing cabinet gone wild. Gray, hulking, and studded with fifty-four dials, the device gauged the brightness of individual stars recorded on photographic plates. Still in use at Caltech, the First Machine fanned Gunn's growing reputation as a plumber—in astronomical parlance, someone with a flair for mechanics of telescopes and the instruments attached to them.

TOUR OF DUTY

Freshly minted Ph.D. in hand, Gunn next fulfilled a military obligation incurred as an ROTC (Reserve Officers Training Corps) student and joined the Army in 1966. As a lieutenant, he completed basic training, then went to paratroop school. Professor Jesse Greenstein, head of astronomy at Caltech, was alarmed at this news. Fearing that Gunn and his talents could be lost to science forever, Greenstein made a few phone calls. As a result, the Army assigned him to the Jet Propulsion Laboratory, where he spent two years helping build cameras that JPL would use for planetary research.

Thereafter, as a researcher shuttling between professorships at Princeton on the East Coast and Caltech on the West, Gunn began to soup up the 200-inch Hale reflecting telescope by rigging it with a variety of exotic accouterments; among these was his first CCD camera for the Hale, the PFUEI. More often than not, his inventions were held together by reinforced plastic packing tape, known around the observatory as Palomar glue. Between astronomical projects, armed with his ubiquitous toolbox, he repaired anything that needed fixing—cars, elevators, computers. "I cannot bear to see something not working," he once admitted.

During the late 1970s and early 1980s, the bearded and balding Gunn spent

James Gunn, astronomer and ace gadgeteer, views the heavens on a video monitor in the observer's cage at the prime focus of the Hale reflector. The image on the screen is provided by the Prime Focus Universal Extragalactic Instrument, a CCD camera of Gunn's invention that is mounted in front of him.

many nights in the Hale observer's cage with PFUEI, often swathed against the cold in a snowmobile suit and almost invariably gobbling M&M's by the fistful while listening to *Rigoletto* or Verdi's *Requiem* on his Sony Walkman. At times like these, he occasionally ruminated on ways that he might further improve on the instruments available for the Hale Telescope. One vexing problem was how to increase a CCD camera's field of view. The thumbnail-size dimensions of the device—its most significant deficiency—provided images that showed only a tiny expanse of the heavens and made even limited sky surveys a tedious process.

In 1978, Gunn came up with a solution, a design for a new instrument to be mounted at the Hale Telescope's Cassegrain focus behind the hole at the center of the 200-inch primary mirror. In essence, his plan envisioned four CCD cameras in one. Light entering the telescope would pass through a window in one end of the instrument. At the other, a four-sided quartz mirror, shaped somewhat like a pyramid, would deflect a different part of the beam through lenses to each of the four cameras, quadrupling the image area of single-CCD cameras. The pyramid design was one that Gunn and some colleagues had devised two years earlier for the Hubble's Wide Field/Planetary Camera.

Besides Gunn, the team that would construct the camera included James Westphal, engineer Michael Carr, and a group of gadgeteers called the Wizards of the Wastebasket, all of whom worked their magic from a maze of cluttered rooms in the basement of a building on Caltech's campus. This crew built the camera on a shoestring budget. Anyone peering into the apparent chaos of its innards would see—in addition to the expected mirrors and lenses—movie projector belts, a razor blade, and foam of the type used in sleeping bag pads, all nestled in a tangle of wires. Gunn even acquired the CCDs, potentially the most expensive of the camera's many parts, by the thriftiest of means: He cajoled NASA into donating four slightly flawed examples left over from building the camera for the Hubble Space Telescope. Perfect ones would have cost more than $100,000 each. In late 1982, after almost four years' work, the camera was finished. Sleek in a white cylindrical sheath, the new instrument was named Four-Shooter for its quartet of CCDs.

To Gunn and Carr fell the task of delivering and installing the precious device, worth about a quarter of a million dollars. With the assistance of others on the team, they loaded the 1,500-pound Four-Shooter into a rented truck early one morning and set out from Pasadena for Palomar Mountain, 120 miles to the south. Carr was so nervous during his stint at the wheel that he slowed to a crawl along the freeway, keeping the speedometer needle pinned to the thirty-eight-mile-per-hour mark. Equally apprehensive, Gunn had a different theory on the reduction of risk: to complete the journey as quickly as possible. He tromped the accelerator and headed for the fast lane as soon as he took the wheel.

Around midafternoon, Gunn maneuvered the truck through a series of hairpin turns leading up the mountain and swung unceremoniously into the parking lot beside the dome of the Hale Telescope. Before dusk the next day, the two men had hoisted Four-Shooter into place at the center of the Hale's vast mirror. After a short shakedown period, the camera began producing images of celestial bodies ghostlier and more distant than any seen before, nearly 12 billion light-years away. And survey work proceeded six times faster with the new instrument than with PFUEI.

Since the completion of the Hale Telescope in 1947, said Gunn, "the push has always been for better detectors. Well, there's nothing on the immediate horizon that will supersede the CCD. It is almost perfect." Thus, by the mid-1980s, astronomy had acquired the countenance of a mature technology: Refinements seemed possible, but no breakthroughs were anticipated.

That appearance, however, would prove to be largely illusion, as innovative telescope builders and mirror crafters began to realize new possibilities for constructing telescopes of greater resolution and light-gathering power than even the mighty Hale.

ELECTRONIC MANIPULATIONS

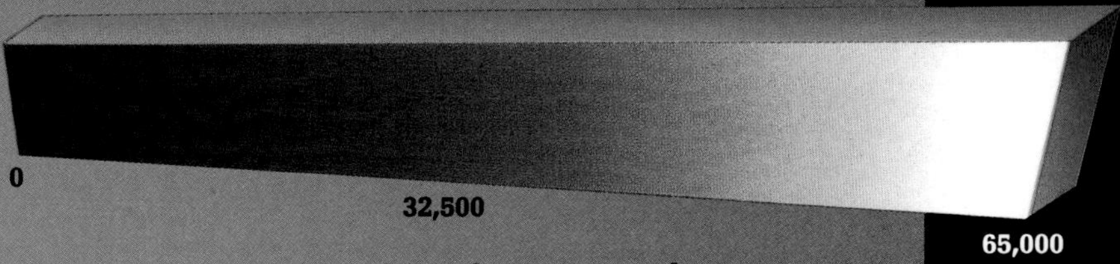

0

32,500

65,000

The light-sensitive electronic chips known as charge-coupled devices have given astronomers a powerful tool in their quest to extract information from the stream of faint radiation reaching their telescopes. CCDs have several advantages over traditional photographic plates. First, they respond to light about seventy times more efficiently than the best photographic film, cutting exposure times from several hours to a few minutes. CCDs are also capable of recording a ten-thousandfold difference in the level of brightness between the dimmest and most brilliant parts of an object, making it possible to capture a galaxy's faint outer regions, for example, without overexposing its radiant center. Equally important, the brightness data gathered by CCDs can be analyzed by a battery of digital image processing techniques. As illustrated on the next two pages, the tiny sensors are designed to produce and store electrical charges when exposed to light. These charges are digitized—or translated into the zeros and ones of binary numbers—for later manipulation by a computer.

Although the process is more difficult, astronomical images acquired photographically can also be digitized, thereby allowing scientists to extract information that may be masked by aspects of the photographic process—emulsion grain, for example—or by limitations of the human eye. In either case, a computer can re-create the varying levels of brightness as shades of gray using a so-called gray scale (above), with black points at 0 and white points at, say, 65,000. The gray scale values can then be adjusted, depending on what aspect of the object is under examination. With image processing, astronomers have been able to look closely at many once-unobservable phenomena in the universe, from the lay of the land on a Uranian moon to the dynamics of stellar evolution. They have also picked up additional clues to some longstanding mysteries about galaxies billions of light-years beyond the Milky Way.

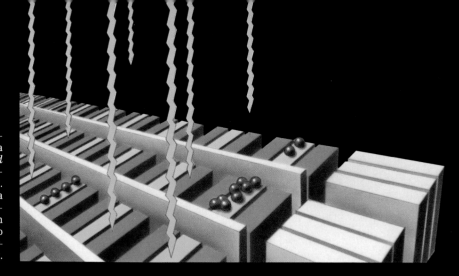

Photons striking a CCD's two-dimensional array of pixels generate a proportionate number of electrons *(red balls)*, which are initially collected in so-called potential wells in each pixel. These wells are electrodes that have a high voltage *(blue)*, which attracts electrons. At the start of an exposure, each high-voltage electrode is flanked by two other electrodes with low voltage *(purple)* to prevent electrons from escaping.

At the end of the exposure, the charges applied to the electrodes in each trio are altered in a controlled sequence. As low-voltage electrodes *(purple)* shift to high *(blue)*, and vice versa, electrons pulled by successive high-voltage electrodes move step by step across the CCD, first within each pixel and then from one pixel to the next.

As the voltage change advances the electrons to another well, a line of pixels is transferred into the "output register," a row of electrodes that is outside the CCD's light-sensitive region and set at right angles to it. When each of the register wells is filled from the adjacent pixel well, electron movement across the CCD comes to a halt.

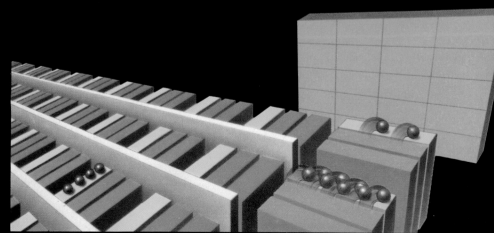

Electrons in the output register are shifted forward one well at a time to a device that amplifies the voltages represented by each group of electrons; another device then converts the voltages into numerical gray scale values. These values will be stored in memory spaces in a computer *(blue grid)* that correspond to the pixels in the CCD.

SILICON MASTERS OF LIGHT DETECTION

Typically, a standard CCD measures light intensity for a million distinct picture elements, or pixels, each representing a small part of the celestial object. The device itself, a wafer-thin chip about the size of a postage stamp, is made of silicon that has been treated so that different areas of the surface possess different electrical properties. As shown here in greatly simplified form, when a pixel is struck by light, it generates an electrical charge that is directly proportional to the intensity of the light that hit it. These charges constitute a latent image, which is electronically processed by a technique known as charge coupling—literally, linking the electrical charge with its corresponding pixel.

Because the light of celestial objects is so faint, the charges are exceedingly weak and must be amplified before being converted to numbers and stored in the computer's memory. Once the brightness recorded at each pixel has been stored, astronomers can program the computer to manipulate that information in a variety of ways, as illustrated on pages 94-99.

Because of their small size, CCDs can record light intensities for only a fraction of a telescope's field of view. CCD images thus are sometimes mosaics made up of several exposures. Depending on the particular CCD and the brightness of the object being observed, several CCD readings can be compiled into a final image in as little as a few minutes.

The first group of electrons is converted to a gray scale value: In this simplified depiction, 2 represents the relatively low intensity of the light that struck the pixel where these electrons originated. The next group to be counted is larger and is assigned a gray scale value of 8. When the output register has been emptied, the next line of electrons will be transferred across the CCD, to the output register, and then on to the computer in a process that will repeat until each pixel has been read and recorded. The varying shades of gray in the computer-generated image at right represent the different numbers of electrons produced by each pixel.

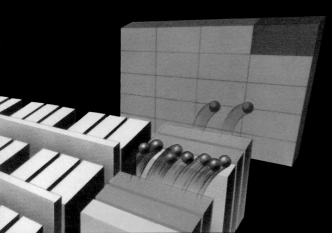

REVEALING HIDDEN FEATURES

With digital image processing, the information gathered in one exposure (whether CCD or photographic) can be rearranged to yield a number of different perspectives on the object in question. In most instances, the manipulations compensate for the limitations of the human eye, which is not suited to detecting subtle differences in brightness. For example, readings for a given portion of an object—such as the cool gas surrounding a hot, bright star—are sometimes tightly clustered in a narrow range somewhere between the two extremes of the gray scale. By designating the darkest and lightest gray points in just the gas cloud itself as 0 and 65,000, respectively, and then redistributing the intermediate values along the new continuum, scientists can heighten the contrast between the cloud and the rest of the image to bring out details of structure that would otherwise be blurred. And since the eye can distinguish only sixteen shades of gray, astronomers often assign colors to the recorded values instead.

Image processing techniques can make use of other kinds of information in addition to brightness levels. By manipulating spectrographic data, for example, astronomers can create images that depict the spatial distribution of the temperature, chemical makeup, and motion of celestial objects—revealing such phenomena as excited hydrogen gas in a galaxy's starforming regions or the varying speeds at which different members of a cluster of galaxies seem to be receding from Earth.

Taken through blue, green, and red filters, three CCD frames were combined to produce an accurate color image *(below)* of NGC 2903, a barred spiral galaxy in the constellation Leo. Each frame was then subjected to a complex process known as maximum entropy technique, which takes into account the blurriness caused by Earth's atmosphere, to yield the image at right, which sharpens details in the galaxy's spiral arms and the dust lanes crossing them.

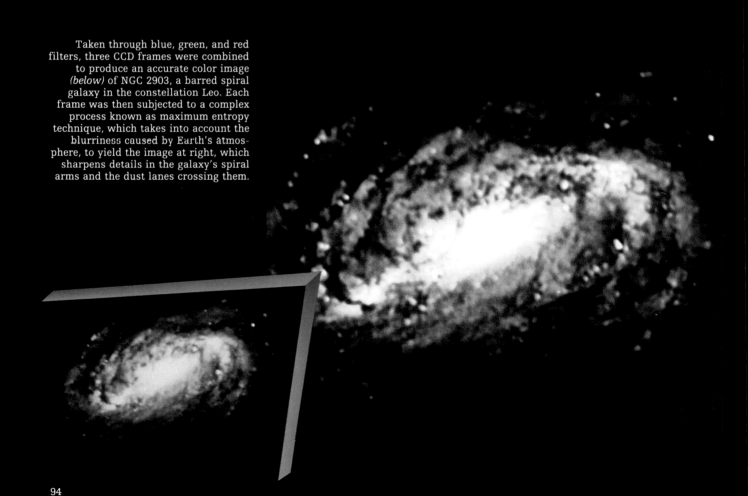

The bright central portion of the Crab nebula, a huge cloud of gas ejected by the supernova recorded in AD 1054 in the constellation Taurus, is mostly featureless in the ordinary photograph below, which is overexposed at the center to capture the faint filaments of its outer regions. At right, a false-color CCD map of the central area—showing the intensity of light emitted by carbon in relation to light given off by hydrogen—reveals areas where carbon is the more plentiful element *(red, orange, yellow)* and areas where hydrogen dominates *(blue, purple)*. The relative abundance of the two elements gives scientists insights into stellar evolution.

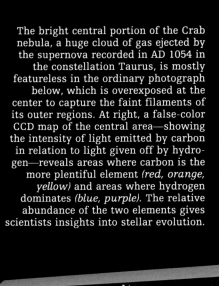

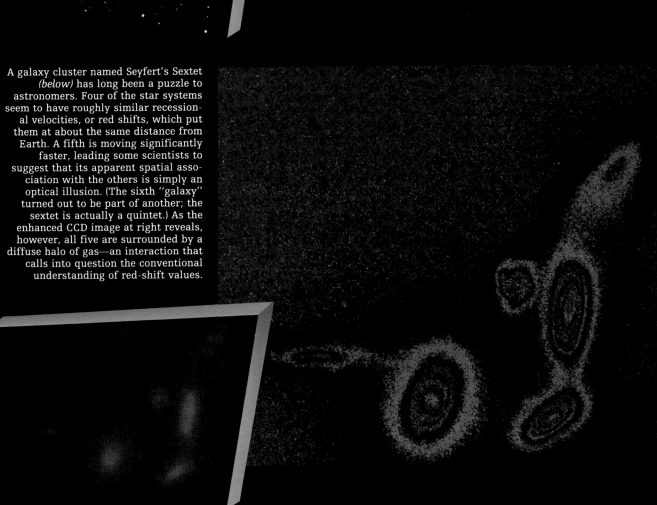

A galaxy cluster named Seyfert's Sextet *(below)* has long been a puzzle to astronomers. Four of the star systems seem to have roughly similar recessional velocities, or red shifts, which put them at about the same distance from Earth. A fifth is moving significantly faster, leading some scientists to suggest that its apparent spatial association with the others is simply an optical illusion. (The sixth "galaxy" turned out to be part of another; the sextet is actually a quintet.) As the enhanced CCD image at right reveals, however, all five are surrounded by a diffuse halo of gas—an interaction that calls into question the conventional understanding of red-shift values.

Spectral analysis and long-exposure photographs like the one below reveal a jet of ionized gas *(red)*—apparently the product of a huge explosion—streaming from the nucleus of galaxy M82 in Ursa Major. Unable to explain the cause of the explosion, some researchers have suggested that the gas jet is an illusion, the result of light scattered earthward by dust. The CCD image at right establishes M82's explosive nature. Made with a special filter, the image shows intense emissions of hydrogen alpha *(black)* emanating from a hot center buried in the galactic disk *(green)*. Scientists now speculate that tidal forces generated by the passage of neighboring spiral galaxy M81 triggered 10 million years of intense star formation in the center of M82, followed by a series of supernova explosions.

The photographic image below of the Ring nebula in the constellation Lyra captures the faint stellar remnant at the center of an expanding ring of hydrogen gas *(red)*. The picture at right is the product of a technique called topographic imaging, which compares adjacent pixels, subtracts their brightness values from one another, and displays the difference. The resulting extreme contrasts between neighboring pixels create shadows and highlights that provide a clearer indication of the spatial relationship between the central star and its ring as well as the varying densities of the ring itself. (Other bright spots are stars in the foreground.)

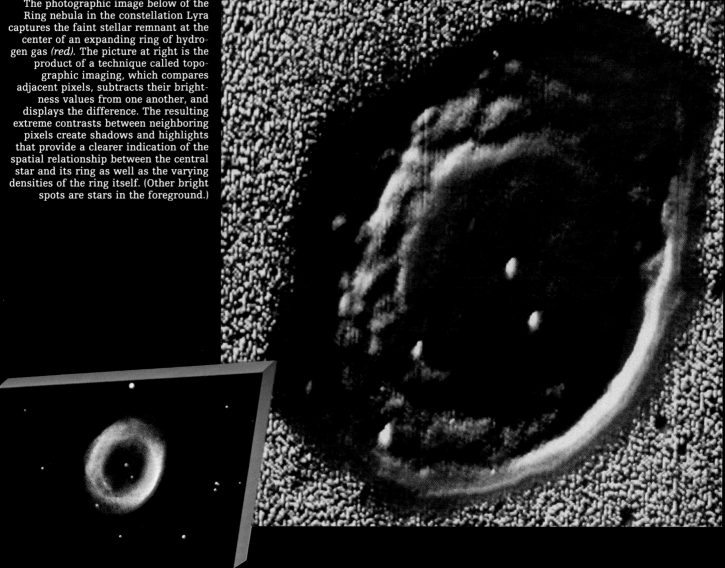

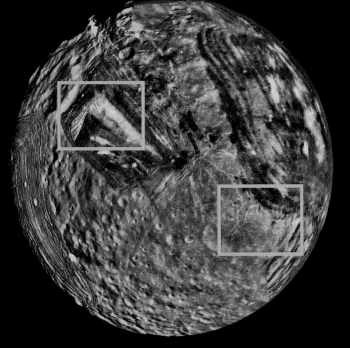

CREATING THE THIRD DIMENSION

On July 20, 1969, when astronaut Neil Armstrong brought the ungainly lunar module *Eagle* in for a landing on the Moon, he had to wrest control from the automatic pilot to avoid coming down in a boulder-strewn crater. All reconnaissance images had depicted the area as relatively smooth, but Armstrong and co-pilot Edwin (Buzz) Aldrin knew they were landing in terra incognita—and experience proved them right.

With powerful computers and digital image proc-

Photographed by *Voyager 2* from 18,000 miles away, the formations in the above mosaic of the south pole region of Miranda, Uranus's smallest major moon, bear witness to the satellite's cataclysmic past. At right is a closeup of the Chevron region *(upper box, above)*. Translated into a three-dimensional perspective view *(below)*, dark areas that appear in the satellite photo to be shallow basins between soft hills emerge as deep valleys alternating with sharp ridges.

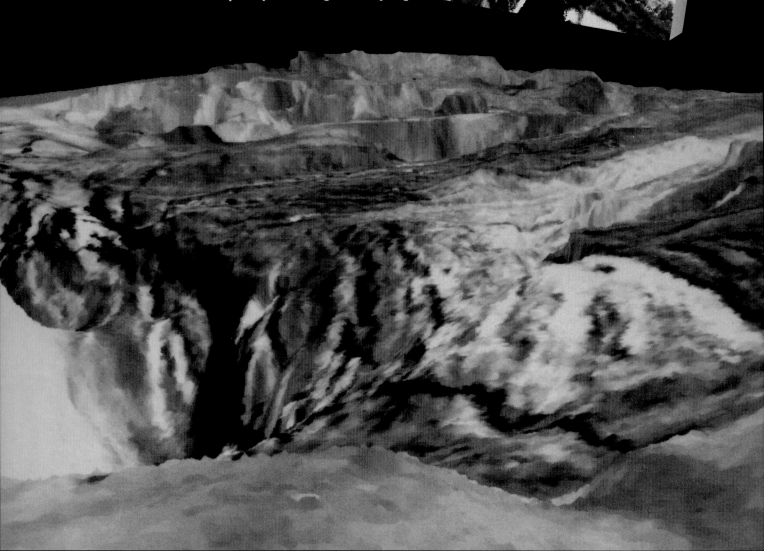

ssing, extraterrestrial landings today, whether phot-ed or robotic, can be plotted more accurately in advance. When a computer is fed gray scale data from two-dimensional satellite images of the surface of a moon or planet, the machine can help earthbound cartographers build a topographic map that shows varying elevations as different shades of gray, and then produce an on-site view as well. The process is similar, in effect, to placing a camera at ground level: The computer is given such variables as the direction the camera will face, its degree of tilt up or down, and its field of view to right and left. By "raising" each pixel in the flat original image by the amount specified on the elevation map, the computer determines the pixel's new location and its gray scale value, thereby creating a three-dimensional image that gives scientists—and prospective human scouts—a better feel for the realities of a proposed landing site.

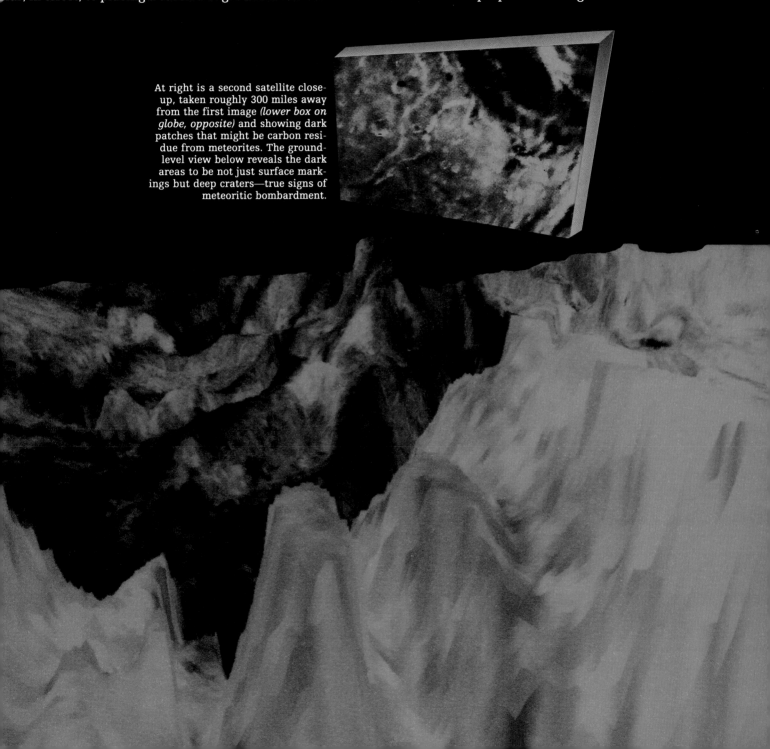

At right is a second satellite close-up, taken roughly 300 miles away from the first image *(lower box on globe, opposite)* and showing dark patches that might be carbon residue from meteorites. The ground-level view below reveals the dark areas to be not just surface markings but deep craters—true signs of meteoritic bombardment.

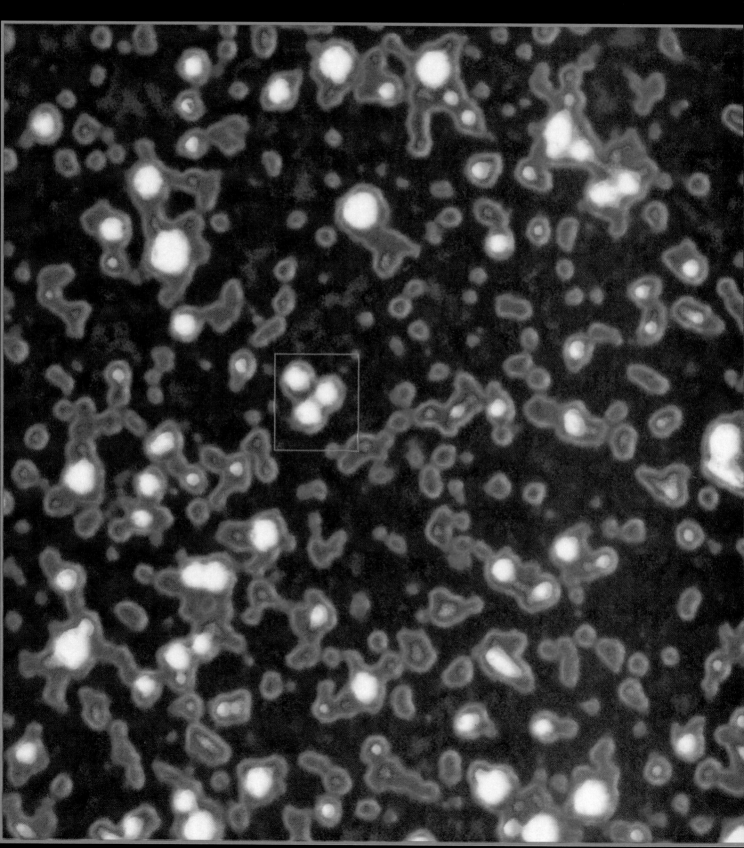

The globular cluster Omega Centauri, a satellite of the Milky Way more than 16,000 light-years from Earth, yields the secrets of its star-packed center in an image captured by the New Technology Telescope in Chile *(inset),* revealing such individual points of light as a triple star system *(box),* shown in greater detail on page 116.

P eaks of the Chilean Andes soar thousands of feet above the layers of atmosphere that most severely distort and dim the spectacle of the heavens. In the thin, frigid air, stars shine strong and bright. To take advantage of these viewing conditions, virtually unexcelled on Earth, roughly a score of telescopes have been built on the Andean heights since the 1960s, and work has begun on a new reflecting telescope that will have a mirror 8 meters across, more than half again as wide as the 200-inch (5-meter) paraboloid nestled at the base of the Hale reflector atop Palomar Mountain. The mirror—so thin that if set on the ground like a bowl, it would collapse under its own weight—will be able to collect more than two and a half times as much light as the Hale Telescope. Some fifty yards away, a twin telescope will rise. By the time the project is completed in 1999, the twins will have become quadruplets, a row of telescopes longer than a football field, with a combined light-gathering power equal to that of a single mirror 16 meters (630 inches) across. Four reflectors in one, the instrument will be named the Very Large Telescope, or VLT.

The VLT, which will permit analysis of spectra gathered from celestial objects more than ten times fainter than those that can be examined with the Hale Telescope, is one of half a dozen or so new and powerful instruments either already scanning the heavens or expected to be in operation by the end of the century. Ingenious elaborations on the basic reflecting-telescope design conceived by Isaac Newton in the seventeenth century, these so-called supertelescopes represent an effort to bypass mirror-making methods that seemed to have reached their limit in the Hale reflector. First, the process of grinding away tons of glass to create a parabolic curve for a conventional mirror bigger than the Hale's would be prohibitively time-consuming and expensive. Second, even if made from a blank like the Hale's—which was engineered with an eye toward saving weight—the finished mirror would have too much bulk. Not only would it be heavy, requiring a costly mount, but also the temperature of the glassy mass would be slow to match that of the nighttime air, an equilibrium necessary to minimize image-destroying convection currents in the air above the mirror. Waiting for the telescope to reach operating temperature would mean losing many precious hours of observation time.

Thus, all of the supertelescopes—whether already completed, under construction, or still in the planning—have primary mirrors that are extraordinarily light for their size. Like the VLT, some of these designs have multiple primaries instead of one. Others have a single composite mirror, a primary built as a mosaic of reflective segments. Some of the mirrors are ground from flat disks, as was the Hale's. Others are cast by spinning molten glass on a turntable to give the mirror a rough approximation of its final, polished contours. And all of these novel telescope designs rely to some degree on guidance from computers to align multiple images, keep each of them in focus, and perform other important functions.

BREAKING NEW GROUND

The first built of these modern reinterpretations of the telescope was the Multiple Mirror Telescope, or MMT. Situated atop Arizona's 8,500-foot-high Mount Hopkins, some forty miles south of Tucson, the instrument consists of six 1.8-meter (70.9-inch) reflectors arranged in a circular array on a single mount. When adjusted so that their images coincide, the six together have the light-gathering power of a single mirror 4.5 meters (177.2 inches) in diameter.

The idea for a telescope of this kind originated with an obscure Irish scientist named Edward Synge. Nephew to playwright John Millington Synge, who was author of *Playboy of the Western World* and other much-lauded works, Edward wrote an article for the August 1930 issue of *Philosophical Magazine and Journal of Science,* a British periodical, suggesting a new design for a telescope. The instrument he had in mind was an "assemblage of a number of similar reflecting telescopes pointed in the same direction, the images from which are ultimately superimposed." Imagining a circular grouping of such instruments, Synge thought of supporting them jointly. He proposed building the support structure of tubes through which air could be circulated to quickly bring the temperature of the mount into equilibrium with that of the surrounding air, thus minimizing the expansion and contraction that would distort the image. Half a century later, in a chronicle of the MMT's evolution, members of the development team noted that their instrument "could be viewed as the realization of the Synge concept," even though MMT designers had no knowledge of his work when informal planning for the telescope got under way in 1967.

At that time, Frank Low of the University of Arizona's Lunar and Planetary Laboratory had begun to think along the same lines as Synge. Low consulted Aden Meinel, an astronomer at the university's Optical Sciences Center and arguably the greatest telescope designer of the 1960s and 1970s, who calculated the optics for such a telescope. Low presented the concept at a NASA conference on large telescopes held in Pasadena in 1969. To that gathering of scientists and engineers, the idea was revolutionary. University of Minnesota astronomer Neville Woolf, for example, remembered feeling entirely leapfrogged by the proposal. "Like many others at the conference," he said, "I was too entrenched" in old ideas about telescopes.

In each of the six tandem telescopes of the MMT, light from distant stars and galaxies would be collected by a primary mirror, which would reflect it to a secondary mirror suspended above it. From there the light would go to a third mirror mounted in the center of the primary mirror. The purpose of the third mirror was to direct the light into a beam combiner—a six-sided conical arrangement of mirrors intended to merge the six light beams together into a single image.

Executing the design for the MMT presented a considerable challenge, especially when it came to combining six images into a single one whose quality would satisfy astronomers. Indeed, the idea seemed so fraught with difficulties that when Low sought help in building the device, he was turned down almost everywhere. One exception was the Smithsonian Institution's Astrophysical Observatory, where, as it happened, scientists had been considering similar proposals. In the early 1970s, the University of Arizona and the Smithsonian agreed to pool their resources. An Arizona team would configure the telescope's optics and arrange for grinding and polishing the mirrors to prescription. The Smithsonian, for its part, undertook the design of the mount and the observatory building.

A TROVE OF MIRROR BLANKS

That Low and Meinel had specified half a dozen telescopes for the MMT was no accident. The U.S. Air Force had planned to use one of the mirrors in a telescope intended for the Manned Orbiting Laboratory (MOL), a research-satellite program abandoned some years earlier. The other six blanks were spares. Meinel sought a donation to the MMT of all seven of the MOL mirror blanks, and the Air Force, having no further use for them, agreed.

Since they had originally been intended for launching into Earth orbit, the hand-me-down blanks weighed only 1,200 pounds apiece. Each was formed with solid front and back plates sandwiching a lightweight eggcrate structure. To reduce changes in shape caused by rising and falling temperatures, the blanks were made of fused silica—a variety of glass having one-sixth the thermal expansion of the Pyrex used in earlier telescopes, including the Hale. In January 1973, technicians at the University of Arizona's Optical Sciences Center began the laborious job of grinding the front surfaces of the blanks into practically identical parabolic mirrors. The MMT team elected to dish them more steeply than usual, decreasing their focal length and thereby the distance from the primary mirrors to the secondary mirrors. The result would be a shorter telescope that would require a smaller, less robust mount and a more compact building to shield the instrument from inclement weather.

To reduce grinding time, the blanks underwent a preliminary process called sagging. Each was heated and its center allowed to sink in order to approximate the required curvature. (One blank melted during this process when it was inadvertently overheated.) Even with this preforming step, however, grinding and polishing the six mirrors would take three years.

Historically, most telescopes used by professional astronomers have been carried in an equatorial mount *(pages 40-41)*, which permits the instruments to track an object across the night sky by rotating on a single axis.

Although this type of mount simplifies the task of following a star in its apparent travels, it also calls for expensive and heavy-duty construction to cradle the telescope.

To lower the cost of the MMT's underpinnings, designers, working under chief engineer for the project Thomas Hoffman and chief scientist Nathaniel Carleton, opted for a so-called "altitude over azimuth" mount. Used with telescopes since the time of Galileo, the alt-az mount served admirably for more than 200 years. Following a star with a telescope so mounted required turning the instrument on two axes, one perpendicular to the surface of the Earth and the other parallel to it. Turning the telescope by hand did not yield precise tracking, but as long as there was only a human observer at the focus, there was no particular advantage in keeping the telescope aimed exactly at a target for long periods, because the eye cannot store light to build an image of a faint star. With the advent of photography in the nineteenth century, however, the alt-az mount was largely abandoned. Photographic film, by accumulating light, revealed objects too dim for the human eye to detect—but only if a telescope gazed unwaveringly at a target for hours at a time. Such precision in tracking was impossible to perform manually and was also beyond the capabilities of mere clockwork mechanisms.

Hoffman and Carleton were able to resurrect the alt-az mount because electronic, digital computers could be programmed to coordinate the rotation around one axis with rotation around the other to keep a telescope trained on its target. And though the required computers were costly, purchasing them was less expensive than constructing an equatorial mount.

Instead of the familiar observatory with a rotating dome atop a stationary base, Smithsonian engineers laid out a squat, rectangular box of a building somewhat resembling a barn. Part of the roof and one side of the building were

A radical departure from conventional telescope design when it was proposed in the early 1970s, the Multiple Mirror Telescope atop 8,550-foot-high Mount Hopkins near Amado, Arizona, has six mirrors, each 1.8 meters (72 inches) wide, arranged around a central computer-controlled beam combiner. The combiner brings the six optical beams to a common focus, yielding an image as bright as that produced by a single reflector 4.5 meters (176 inches) in diameter.

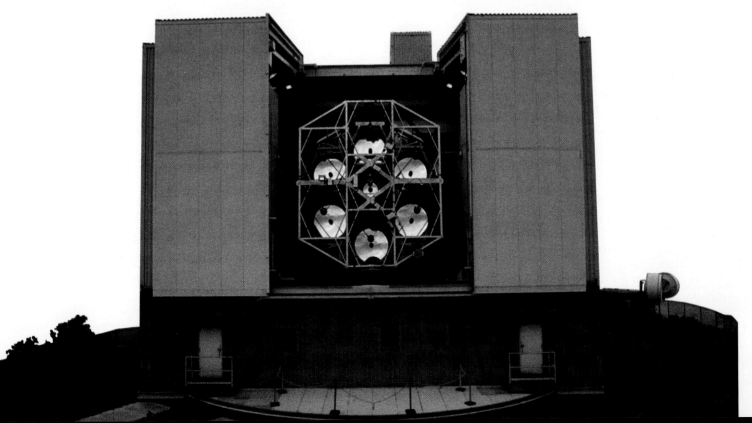

designed to slide open during observations, and the entire 450-ton structure would turn slowly on enormous steel wheels to match the speed of the telescope as it tracked a star or other heavenly body. The squarish building would cost less to build than a more conventional observatory offering a like amount of space for laboratories and other work areas.

In May 1978, the MMT's six mirrors, polished and coated with a reflective layer of aluminum, were set into the mount, and several weeks were spent aligning them. Then came the time to try combining the six images into one, a process called coalignment. The approach adopted for the MMT involved the use of a bright guide star near the subject of an evening's observations. All six telescopes were aimed at the guide star. Next, a telescope operator initiated the coalignment process—called stacking the images—by manually adjusting the secondary mirrors to move the six images into a rough stack. From there, the computer took over, using cleverly conceived algorithms first to determine the exact center of each image and then to superimpose them on one another. Thereafter, the computer automatically adjusted the mirrors according to rules that account for subtle warping by gravity as the telescope tilts to track a star. Because this distortion is complex, the rules describing it are imperfect at best, and the coalignment process must begin anew every fifteen minutes or so.

OFF TO A GOOD START
Even before the formal dedication of the Multiple Mirror Telescope, which took place during the month of May, 1979, the instrument was producing images of excellent quality. During its very first observation, it gathered evidence confirming the discovery of a so-called gravitational lens, made with the Hale Telescope a short time earlier. Even so, the new telescope was not performing quite as well as its designers thought it could. Part of the problem lay with the observatory floor, which was incomplete when the telescope went into operation. Holes admitted drafts that disturbed the air in front of the instrument, slightly degrading its resolution. This source of thermal disturbances was quickly and easily eliminated.

Other perturbations of air around the telescope arose from the mount itself, which absorbed heat during the day and released it during the cool of the evening, causing subtle convection currents. Plans for the MMT had called for a powerful air-conditioning system to keep the temperature of the telescope near the temperature expected during the night, but other, more important features of the observatory consumed the money set aside for the cooling equipment.

To minimize the problem, astronomer Jacques Beckers, the director of the observatory, first set his staff to photographing the entire building with an electronic camera that was sensitive to infrared radiation, the wavelength of light emitted by all warm objects. In the resulting images, warm objects showed up as brighter than cool ones, thereby pinpointing hot spots on the observatory floor and ceiling, as well as on the telescope structure itself. The

THE HUBBLE: FLAWED BUT STILL PHENOMENAL

High above Earth's atmosphere, the space shuttle *Discovery* released the Hubble Space Telescope into orbit on April 25, 1990 *(below)*. The $1.6 billion craft was expected to reveal celestial objects at a resolution ten times greater than astronomers can achieve with ground-based instruments.

Such expectations vanished, however, when the Hubble's first views proved to be surrounded by fuzzy halos, the result of a two-micron error—one-fiftieth the width of a human hair—in the telescope's primary mirror. Luckily, the malady is at least partially curable. Computer processing can be used to sharpen many images, and in 1993, shuttle astronauts will be sent to fit the telescope with corrective mirrors. Despite the Hubble's travails, its 380-mile-high perch provides a breathtaking view of the Solar System and the universe beyond, as demonstrated by the photographs on the following pages.

staff responded by insulating the floor with polystyrene and carpeting. They caulked the tiniest chink in the observatory walls and wrapped reflective insulating tape around much of the telescope mount to keep its temperature from varying so much between day and night. These homely remedies worked so well that when the atmosphere above Mount Hopkins was steady, the Multiple Mirror Telescope produced images up to 20 percent sharper than it had before being insulated.

Together, the MMT's sextet of mirrors collect about 80 percent as much light as the Hale Telescope. When enhanced by speckle interferometry techniques *(pages 78-79)*, the resolution that can be achieved with the MMT's six

Caught in an intimate gravitational dance 3 billion miles from Earth, Pluto and its moon Charon look like siamese twins in even the best ground-based photograph *(near right)*. The unprecedented view afforded by the Hubble *(far right)* reveals the duo as distinctly separate worlds for the first time.

mirrors is about one-third better than that obtainable with the Hale using similar methods. By any measure, the Multiple Mirror Telescope has been a successful test of a new technology.

TO OUTSHINE THE HALE

In 1977, as work was being completed on the observatory building and mount for the MMT, Robert Kraft, director of Lick Observatory headquarters at the University of California's Santa Cruz campus, pondered the school's future in astronomy. Lick Observatory, with its 120-inch reflector near San Jose, was in danger of becoming obsolete. In the years ahead, increased light pollution

Two photographs of the same region of M14, a globular
cluster of stars located 70,000 light-years from Earth, illus-
trate the advantage of the Hubble's lofty perch. The fuzzy
blobs in the ground-based image *(above, left)* resolve to
hundreds of distinct pinpoints of light when viewed by the
Hubble *(above, right)*. Astronomers had turned the telescope
toward M14 in search of a star that had brightened into a
nova in 1938. The Hubble's superior vision revealed that what
was thought to be a single likely candidate was in fact a
group of half a dozen stars.

from nearby communities was expected to turn the observatory into a second-
rate facility. The university, which had long enjoyed a position at the forefront
of astronomy, needed a new and better telescope.

To address the issue, Kraft formed a committee. One of the members was
Jerry Nelson, a research fellow at Lawrence Berkeley Laboratory. (The lab is
a federal facility administered by the university.) While others evaluated the
MMT and single-mirror approaches to building a telescope, Nelson investigated an alternative design in which a single primary mirror would be produced in segments that fit together like pieces of a jigsaw puzzle. After two
years, he and his superiors decided in favor of a segmented mirror.

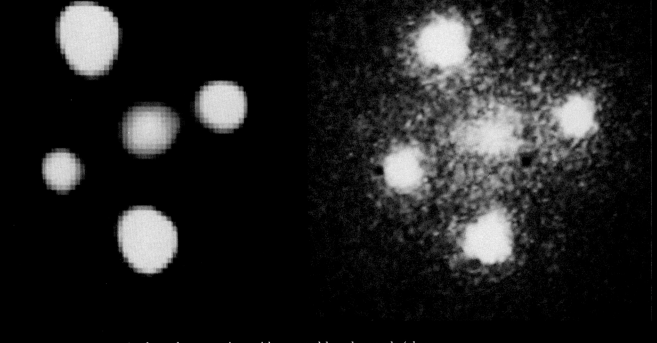

As shown by comparison with a ground-based example *(above, left)*, the Hubble photograph of a quintet of bright lights named G2237 + 0305 *(above, right)* is the most detailed image ever made of the object—a gravitational lens system also known as the Einstein Cross. According to Einstein's general theory of relativity, gravity warps space, thus bending the path of passing light. In the case of the Einstein Cross, radiation emitted by a single quasar 8 billion light-years away is bent by the mass of an intervening galaxy, so that four quasars appear to surround the galactic core.

The idea of splitting a primary mirror into pieces seems to have originated with Lord Rosse, the nineteenth-century Irish peer who created the 72-inch Leviathan of Parsonstown, the largest reflecting telescope ever built with a metal mirror. As early as 1828, Lord Rosse performed an experiment with a two-piece composite mirror. In search of a way to reduce spherical aberration in telescope mirrors without troubling to grind them as paraboloids, Lord Rosse cut the center out of a spherical mirror, lowered this central portion within the outer ring of the original mirror, and supported it from below with screws. The result was a two-step approximation of a paraboloid.

Lord Rosse did not pursue his invention beyond the experimental stage, but the notion of spherical mirror segments arranged as a paraboloid surfaced again more than a century later. In 1932, as a result of writing an article about the building of the Hale 200-inch, Italian astronomer Guido Horn-d'Arturo was inspired to design a single large reflecting surface as a mosaic of smaller ones, rather like a tiled birdbath. Soon he began construction of just such a mirror, using tiles of spherically ground glass, in the tower of the Bologna Observatory, of which he was then director. Without much financial support, the enterprise did not progress very rapidly and was further set back by the outbreak of World War II. Horn-d'Arturo resumed his work in the late 1940s, and by 1953 he had created a composite mirror that consisted of sixty-one tiles measuring 20 centimeters (7.8 inches) in diameter and arranged in four concentric rings. Each tile was adjusted by means of three screws to give the reflecting surface the shape of a paraboloid and focus the component images at a single point. Together, they were equivalent to a single mirror 1.8 meters (70.9 inches) in diameter. The telescope produced acceptable photographs, but it made no significant contribution to the science of astronomy.

A FEATHER FOR NELSON'S CAP

The composite mirror envisioned by the University of California's astronomy committee would be far larger—10 meters in diameter, double the size of the Hale mirror. Caltech agreed to join the effort in order to spread the cost of the telescope, creating a group named the California Association for Research in Astronomy. As it turned out, the project was largely funded by a $70 million gift from a philanthropic foundation established by the late William M. Keck, a wealthy oil wildcatter who had served on Caltech's board of trustees. The grant stipulated a name for the instrument: the W. M. Keck Telescope. A site on Mauna Kea in Hawaii was selected for this new eye on the universe.

The design for the Keck Telescope called for its expanse of mirror to be divided into thirty-six hexagonal segments, each 1.8 meters across. The segments would fit closely together in three concentric rings to form a hyperbolic reflector. A hexagonally shaped hole in the center would allow light to be directed to a Cassegrain focus behind the mirror.

The advantages of a segmented mirror were as obvious to the Keck designers (by now, Jerry Nelson was chief scientist of the project) as they had been five decades earlier to Horn-d'Arturo. Weighing only 880 pounds each, the small-diameter segments could be just 7.6 centimeters (3 inches) thick without risking the loss of image quality from sagging. The primary's relatively light weight, its short focal length, and the use of an alt-az mount having considerably less mass than the equatorial variety all helped keep the cost of the Keck Telescope down. Another advantage of the composite design was that it would make the periodic renewal of the mirror's reflective aluminum coating less risky. If one segment were to break when it was removed from the mount for recoating, it could be replaced without great difficulty.

Nelson and his team began work on the Keck in 1980. A decade later, their labors were almost complete. The observatory building, which rests on the flattened top of a cinder cone near the volcano's summit, was finished. Inside the dome—a huge white bubble against an azure sky—nine of the telescope's three dozen planned mirror segments were about to be put in position on the mount. Even with only one-fourth of its mirror surface in place, the Keck would equal the Hale in light-gathering ability.

THE INFLUENCE OF GRAVITY

As is the case with any telescope, the finished composite mirror will be subjected to the varying influence of gravity as the telescope raises or lowers its gaze to track a target, to buffeting by wind, and to warpage caused by changes in temperature. Maintaining the proper alignment of all the segments to within millionths of an inch, as required for sharp imaging, is a feat made possible by a so-called active-alignment system. Positioned between the mirror segments are 168 sensors, each able to discern misalignments a thousand times smaller than the width of a human hair. The sensors pass position data to a computer, which in turn sends instructions to 108 actuators, each of which controls a small motor that turns a tiny precision screw to push or pull a segment into place.

The telescope is expected to be especially valuable in spectroscopic studies of stars and galaxies. The Keck will be able to gather enough light to capture the spectra of objects four times fainter than can be examined with the Hale. Says Nelson, "We'll be able to look back a long way, to a time when the universe was very different from what it is now."

Meter for meter of primary mirror diameter, the Keck Telescope promises to be one of the most economical telescopes ever built, but it will soon be eclipsed in that respect by an instrument called the Spectroscopic Survey Telescope, or SST, located at the McDonald Observatory in Fort Davis, Texas. To be constructed in a building that resembles a grain silo, it is intended to be used exclusively for spectroscopic study of stars and galaxies. A joint project of the University of Texas at Austin and Pennsylvania State University, the SST was conceived in 1983 by Penn State astronomer Daniel Weedman. When finished, perhaps as early as 1994, the entire telescope will have cost about $9 million, approximately one-tenth the amount spent on the Keck.

PROFILE OF A BARGAIN

The main economies of the SST come from its fixed tilt. Because the gaze of the telescope is permanently set at 60 degrees above the horizon, the effect of gravity will remain constant, eliminating the need for costly mechanisms to realign the mirrors. But the instrument is mounted on a turntable, enabling it to cover a strip of sky 72 degrees wide by rotating. In comparison, a telescope that can change its angle of elevation has access to a strip of sky more than 90 degrees wide, but the view of the SST is amply broad, says Weedman, "to keep us busy."

Additional savings accrue from the mirrors chosen for the SST. Instead of tricky-to-grind parabolic or hyperbolic segments, the SST's composite pri-

A little-known forerunner of modern segmented telescope mirrors was the mosaic-like reflector conceived in 1932 by Italian astronomer Guido Horn-d'Arturo and built by him two decades later in the Bologna Observatory. Horn-d'Arturo arranged sixty-one hexagonal tiles of identical spherical curvature in four concentric rings, thus forming a shallow paraboloid without having to grind that difficult shape. Screws attached to the underside of each tile allowed the astronomer to tilt the mirrors individually, channeling their light to a common focus and eliminating spherical and other types of optical aberrations.

mary uses eighty-five identical, spherical mirrors 1 meter (39.4 inches) in diameter. These are arranged nearly touching one another in a steel-rod framework that has a curvature just matching the arc of the mirrors. The result is a spherical reflecting surface about 8 meters (315 inches) in diameter. Because the chief drawback of a spherical mirror is its inability to focus light as sharply as the parabolic or hyperbolic variety can, the design team for the SST invented a special device to be mounted above the primary mirror. It contains a system of mirrors that compensates for the primary's spherical aberration *(pages 126-127)*.

The SST will have two of these devices. They are called trackers, a name that reveals their second crucial function: to enable the immobile telescope to follow stars across the sky as it spins with the Earth. Each tracker is to move along precise tracks in the mirror's focal plane, the imaginary surface above the mirror on which objects are most sharply focused. Directed by computers, the trackers can keep two targets in view simultaneously for up to an hour—long enough, Weedman anticipates, to record the spectra of a multitude of stars and galaxies that have not yet been studied spectroscopically.

SPUN FROM MOLTEN GLASS

While designers like Nelson and Weedman sought to make a big mirror from smaller ones, an Oxford-educated physicist named Roger Angel decided to take a more direct approach. His goal was to figure out a way of making one-piece mirrors with diameters as large as the SST's composite reflector at a fraction of the cost of the Hale mirror. Angel arrived at Columbia University in 1967 to begin postdoctoral studies in physics, with no intention of becoming a telescope maker. But his faculty adviser was involved in making telescopes used for studies in the x-ray region of the electromagnetic spectrum, and Angel not only became deeply engaged in that subject but acquired a fascination with astronomy in general. In 1973, he moved to Tucson and a position first as an associate and then as a full professor of astronomy at the University of Arizona's Steward Observatory. By the early 1980s, Angel had become determined to construct a very large diameter, one-piece mirror of the kind the experts had declared both too difficult and too expensive to build.

Angel saw a solution to the weight and expense of casting a large mirror in the simple expedient of spinning the circular mold used to shape a mirror blank. He claims no credit for the concept, which he says "is as old as the hills." The technique had already been employed in the early 1900s to spin a dish of mercury into a paraboloid reflector. (It could only look straight up.)

When a liquid spins in a vessel, forces generated by the rotation cause the fluid to flow from the center toward the edge. The result is a nearly perfect parabolic surface. In Angel's mirror-making application, such a surface, formed of molten glass that was subsequently allowed to cool and harden as the mold continued to spin, would need a minimum of grinding before it could be polished. Almost as a bonus, this approach could also be used to make reflectors of more pronounced curvature (shorter focal length) than their predecessors simply by spinning the mold faster. To minimize weight without sacrificing rigidity in the finished mirror, Angel hit upon an idea not yet used in spun mirrors. He decided to structure the blank as a honeycomb, leaving only the face solid for polishing. Together, short focal length and light weight would permit some of the most economical telescope mounts ever built.

With no experience in glassworking, Angel had to begin from scratch. And rather than seek formal training in the craft, he began to teach himself, melting old Pyrex cups and bowls in a backyard furnace just to get a feel for the materials. In the mid-1980s, he established the Steward Observatory Mirror Laboratory, then proceeded to build his first rotating oven in a former synagogue in Tucson. Mirror-making work began modestly with a 1.8-meter (70.9-inch) mirror. Angel, graduate student John Hill, and a small engineering

University of Arizona telescope designer Roger Angel, inventor of the spin-cast mirror, holds a paper mock-up of the honeycomb core that has allowed him to build reflectors up to 3.5 meters (138 inches) in diameter. Made of alumina silica fibers, the core is flushed out after the spun mirror blank cools, leaving the honeycomb structure that gives the mirror its trademark lightness and strength.

staff started by securing hexagonal ceramic blocks some 20 centimeters (7 inches) across and about 30.5 centimeters (12 inches) tall to the bottom of a mold, shaped like a large cake pan and set on a turntable. Then they gently added, through the open top of the oven, chunks of a Pyrex-like formulation of borosilicate glass from Japan.

Next, a crane lowered a lid onto the oven, which was then turned on. Supplied with seventy-five kilowatts of electricity—power enough to supply 750 hundred-watt light bulbs—passed through a heating element similar to those found in common toasters, the temperature of the oven steadily rose to about 2,000 degrees Fahrenheit. Over a period of twelve hours or so, the glass melted, filling in the spaces between the ceramic hexagons and forming a layer of glass above them little more than an inch thick. Then, the entire sixteen-foot oven began to spin, accelerating to sixteen revolutions per minute, a speed determined by the curvature desired for the mirror. Gradually, the liquid glass—a bright cherry red—spread into the parabolic shape that Angel had planned for the mirror.

A SLOW COOL-DOWN

The melt was monitored by television cameras. After any bubbles had risen to the top and burst, and when the surface of the molten glass appeared free of other blemishes, the temperature inside the turning oven was slowly reduced. After several weeks of cooling, at the rate of about one-quarter degree per hour, the mirror reached room temperature. Technicians then played a stream of high-pressure water on the hexagonal blocks, now crumbly from baking, to wash them away, and removed the mold. The result was a parabolic mirror blank light enough to float facedown on water, backed by a rigid honeycomb that permitted circulating air to quickly equalize the temperature of the glass and the temperature of the surrounding atmosphere.

The final step, polishing the mirror, posed additional problems. The traditional tool for this job is a spinning, disk-shaped tool called a lap. For shallowly dished parabolic mirrors, in which the curvature changes only slightly between the center of the reflector and the edge, the lap can be nearly as wide as the diameter of the reflecting surface itself. But for Angel's mirror, in which the difference in curvature between center and edge was substantial, a traditionally designed lap would have to be no more than a couple of inches across. Exaggerating somewhat, Angel likened the polishing of his mirror with a two-inch lap to "cleaning the Washington Monument with a toothbrush." To speed the process, University of Arizona staff scientist Buddy Martin came up with a new design: a flexible lap whose shape is adjusted 1,000 times per second by a system of computer-controlled, motor-driven levers set around its circumference. Although Martin's system permits a tool that is two feet in diameter, polishing is still expected to take a full thirteen months. The blank will then be coated with a reflective layer of aluminum before it is installed, sometime during 1992, in a telescope being built by the Vatican Observatory atop Mount Graham in Arizona. (The Vatican established an observatory in the sixteenth century at Castel Gandolfo, the summer residence of popes.)

Though serviceable, the oven that Angel used to cast the 1.8-meter mirror was less than ideal: To save money, controls for the oven and the instruments for monitoring the progress of the spin-casting process had been installed on the turntable. Operating the device, therefore, required the services of an "oven pilot" to ride the turntable and keep close watch on the temperature, bubbles in the melt, and other factors. Most who filled the position succumbed to dizziness and nausea. So, in 1986, when Angel built a much larger oven beneath the grandstand at the University of Arizona football stadium, he installed the controls in an adjacent room. The new oven was equipped with a 12-meter (39.4-foot) turntable that can accommodate a mold capable of holding eighteen tons of glass. Two years later, it was used to spin-cast three 3.5-meter (138-inch) mirrors, one of which was destined for a telescope at the Astrophysical Research Consortium's Apache Point Observatory in New Mexico, one for the National Optical Astronomy Observatories at Kitt Peak in Arizona, and the third for the U.S. Air Force to use in experiments in adaptive optics *(pages 122-123)*.

SIX METERS AND BEYOND

Coming up on Angel's agenda is a 6.5-meter (256-inch) mirror commissioned to replace the six mirrors of the MMT on Mount Hopkins. The multiple-mirror layout for that telescope evolved at a time when there was no way to acquire comparable light-gathering power economically with a single mirror, a technological vacuum filled by Angel's spin-casting technique. The new mirror will fit the existing mount with a minimum of modification, and when it is installed in 1994, it will not only more than double the facility's light-gathering power but also make the telescope more convenient to use by ending the need to stack the instrument's six component images. That the MMT is abandoning its six individual telescopes, however, is not condemnation of the multiple-mirror approach. Indeed, a number of supertelescope designs follow the same philosophy, and for good reason: A telescope can always see more with multiple mirrors of the largest economically practical size than it can with just one.

One example is the Columbus Project Telescope, planned to enter service in the mid-1990s atop Arizona's Mount Graham near the Vatican Observatory's new telescope. The design of the Columbus—a cooperative effort between the University of Arizona and Italy's Osservatorio Astrofisico di Arcetri in Florence—calls for a matched pair of 8-meter (315-inch) mirrors to be cast by Angel and the Steward Mirror Laboratory. The first of these behemoths is scheduled for casting in the early 1990s; the other will follow a year or so later. With two separate optical systems yoked together on a single mount, the telescope, which is also known as Two-Shooter, will have light-gathering power equivalent to that of an instrument having a single mirror 11.3 meters (445 inches) in diameter.

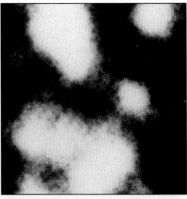

1 meter: 2 arc seconds

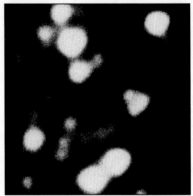

3.6 meters: 1 arc second

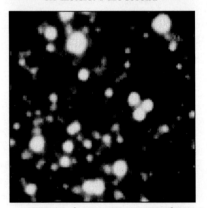

NTT raw image: .33 arc second

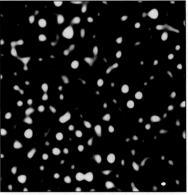

NTT processed image: .18 arc second

THE TEUTONIC ALTERNATIVE

But Roger Angel no longer holds a monopoly on lightweight optics for reflectors. The German firm Schott Glaswerke in Mainz, for example, casts mirrors that resemble colossal contact lenses. With a curved rear surface as well as a parabolic face, these mirrors are called meniscus mirrors, named for the crescent shape taken on by the surface of a liquid standing in a narrow container, such as a test tube.

Schott uses one of two mirror-making techniques, depending on the size of the blank required. For mirrors up to 4.5 meters (173.2 inches) in diameter, Schott simply cuts a slice from a cylindrical plug of glass, then grinds and polishes this blank to its final meniscus shape. Such was the procedure followed in making a 3.6-meter (142-inch) mirror for the New Technology Telescope, or NTT, which recorded its first image—of the globular star cluster Omega Centauri—from a mountaintop near La Silla, Chile, in 1989. This instrument is the creation of the European Southern Observatory (ESO), a consortium of European nations formed in 1962 that includes Sweden, West Germany, Belgium, the Netherlands, France, Denmark, Italy, and Switzerland. With headquarters near Munich, ESO operates fifteen telescopes at its site in northern Chile, 375 miles north of Santiago.

The NTT mirror is only nine and a half inches thick at any point. Without a honeycomb or other structure to impart rigidity, this thin, elegant reflector could not possibly hold its shape were it not for a sophisticated active-optics system designed by ESO's optical wizard Ray Wilson. It consists of seventy-eight computer-controlled levers that both support the mirror and subtly bend it to peak optical profile against stresses caused by gravity, wind, and variations in temperature. Because a meniscus mirror is so flexible, compared with other varieties, it adapts quickly to the computer-directed fingers; errors in curvature disappear almost magically. And because there is so little glass in the meniscus, it is quick to reach thermal equilibrium. Weighing even less than a mirror spin-cast at the Steward Mirror Laboratory, a meniscus reflector requires a relatively light structure to support it, reducing the cost of building a telescope mount.

THE ADVENT OF ACTIVE OPTICS

Although praiseworthy in its own right, the NTT mirror is most important as a trial of the active-optics system needed for the Very Large Telescope. Each of the VLT's four 8-meter meniscus mirrors, ordered from Schott Glaswerke in late 1988, is only seven inches thick and must be supported on more than 150 contour-molding levers. Because of their larger diameter, these mirrors are spin-cast rather than poured into a stationary mold. The procedure, however, differs from Angel's. Instead of melting the glass in the mold, which is then spun, the German firm melts the glass in a separate crucible, then pours the liquid into a mold that is already in motion. The first mirror is to be completed in time for a 1995 installation date.

Conceived as an array of four individual telescopes arranged in a line, the

Although resolution improves somewhat between the first and second photographs at left, made with older telescopes at the European Southern Observatory in Chile, a triple star system in Omega Centauri appears at best as a roughly triangular blur of light. The stars only begin to shine individually in the third picture, a CCD image made through the ESO's New Technology Telescope. The 3.6-meter (142-inch) reflector's focus is controlled by a system of active optics designed by chief optician Ray Wilson *(below)*. In the final image, enhanced by computer processing, the angular separation between the stars is .18 arc second—equal to picking out a dime at a distance of fifteen miles.

VLT is intended to be the most capable and most versatile telescope ever built. For example, each mirror will be able to look at a different object, or all four can observe the same object but at different wavelengths. Or the telescopes can be aimed at the same faint target and their images combined to make it brighter. This approach promises unprecedented light-gathering power. Together the four 8-meter mirrors will be able to collect as many photons as a single mirror 16 meters across. Through a technique called long-baseline interferometry, widely used by radio astronomers since the late 1950s, ESO also hopes to achieve the resolution that is theoretically possible with an instrument that has a mirror 100 meters in diameter. Such acuity would advance the study of preplanetary nebulae—clouds of gas and dust thought to be condensing into solar systems in orbit around distant stars—and would help astronomers discern heretofore invisible detail in the centers of galaxies.

The VLT and its 8-meter spin-cast mirrors seem to lie on the edge of what can be achieved with optical telescopes. Astronomers can only wonder what they will discover with this marvelous new instrument, the Keck Telescope, the SST, and others. Since Galileo's time, steady advances in the quality of telescopes have regularly upset humankind's view of the universe. Visible light, focused through ever sharper and brighter telescopes, revealed, for example, that millions of celestial bodies thought to be nebular clouds of gas and dust are in reality swarms of stars on a galactic scale. Paired with photography in the nineteenth century—and later with electronic detectors so sensitive that they can capture individual photons of light—telescopes have maintained their preeminence as the medium through which astronomers gain most of their knowledge of the cosmos. With the powerful new instruments under construction and on the drawing boards, astronomers may be rewarded with their clearest glimpse so far at the birthplaces of stars and the centers of galaxies and perhaps the discovery of planets orbiting other suns, too dimly lit to have been seen by lesser telescopes. In greater detail than ever before, they will be able to study the universe as it was when it was new. And they may be afforded sights never even guessed at. As Keck Telescope designer Jerry Nelson says, "Probably the most exciting things we'll discover are the things we haven't thought of yet."

A NEW GENERATION OF MIRRORS

For three decades after the Hale 200-inch mirror went into operation, the building of large mirrors for ground-based observation seemed to have reached a point of diminishing returns. Although a mirror twice the diameter of the Hale would collect four times as much light, the quality of its images would depend on its ability to hold its shape precisely. To be rigid enough, a traditional large mirror would have to be quite thick—an increase in mass that would have two adverse side effects. The heavier the mirror, the more it would tend to sag—and sag in different ways as it was repositioned to point at different objects. Thick glass would also cool more slowly than the night air, causing a temperature differential that would create image-distorting eddies and currents above the mirror's surface. In addition, the differential temperature within the glass itself would warp the mirror. Although the designers of the Hale effectively address-ed both of these problems *(pages 60-61)*, the mirror took two decades to build. Moreover, the costs associated with large mirrors —including the massive structures to support them—tend to rise as a cube of the diameter: A mirror twice as big as the Hale would cost eight times as much.

Recently, however, radical departures from traditional design, cou-pled with such technological advances as computerized controls, have begun to offer a way around the problem of building large mirrors. As illustrated by the telescopes shown on the following pages, some solutions incorporate systems that compensate rapidly and automat-ically for distortions caused by gravity, wind, or atmosphere. Others drastically reduce the time needed to grind, polish, and shape huge mirrors by constructing a composite primary or using new techniques for casting the mirror blank. These revolutionary telescopes are scheduled to see first light in the 1990s, and all promise ground-based astronomers their deepest views yet into the universe.

Out of a Spinning Furnace

An innovative process called spin-casting, developed by astronomer-designer Roger Angel of the University of Arizona, has opened the way to making individual mirrors larger than the Hale 200-inch in a fraction of the time and at a fraction of the cost. Angel has already produced smaller reflectors by this method, with the aim of building 8-meter (315-inch) versions for a two-mirror telescope known as the Columbus. Scheduled for installation in the mid-1990s atop Mount Graham, about 100 miles northeast of Tucson, the Columbus is a cooperative project involving members of the astronomical communities of the United States and Italy. Together, the pair of mirrors will equal the light-gathering capacity of a single 11.3-meter (445-inch) mirror—five times the capacity of the Hale.

Following in the Hale's venerable footsteps, the Columbus reflectors will incorporate a honeycombed interior to reduce weight without sacrificing rigidity, a combination that will make the mirrors resistant to gravitational distortion. The design also enables the mirrors to reach thermal equilibrium even more quickly than the Hale, saving valuable observing time. However, spin-casting the mirror blanks will save years in achieving the final product.

The process begins when chunks of borosilicate glass are fed into a mold set inside a large, rotating oven *(right)*. As the glass melts, the spinning forces

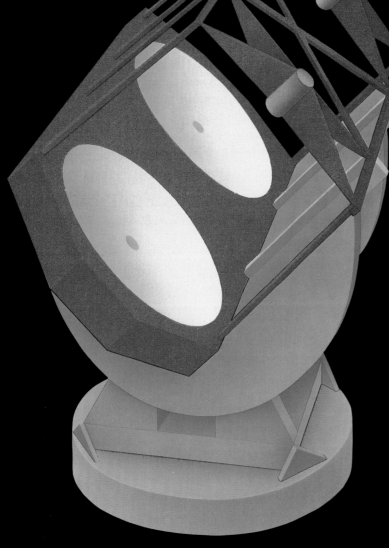

With its mirrors yoked together on a common mount, the Columbus will resemble a giant pair of binoculars. Light reflected from the primary mirrors is sent to secondary mirrors 8.5 meters (28 feet) above the surface of each primary, yielding a compact structure that minimizes wind vibrations and tracks easily.

a large proportion of the molten material toward the edges of the mold, producing a natural bowl shape. The casting process thus performs much of the work of shaping the mirror: Rather than having to grind away several tons of glass from a flat disk to form a parabolic shape—a difficult and tedious process that required several years in the case of the Hale—technicians can skip the rough grinding stage and go straight to polishing the mirror blank.

Another important benefit of spin-casting is that the parabola can be made deeper simply by spinning the furnace faster. The more steeply curved the parabola, the shorter the focal length of the mirror and the more compact the telescope. Prime focus for the Hale, with its shallow parabola, is 16.8 meters (55 feet) away from the mirror surface, requiring a cannonlike tube. For the Columbus mirrors, prime focus will be only 9.6 meters, or 32 feet, away, a squat configuration that offers less of a target for buffeting by wind and requires much less in the way of expensive protective housing.

With its tremendous light-gathering capacity, the Columbus will be used primarily for spectroscopy on very faint celestial objects. Each of the mirrors can be separately instrumented and their light subsequently combined by computer to avoid the dimming that would result if extra mirrors were used to combine light from the two primaries.

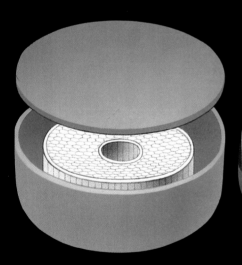

The raised lid of the spin-casting furnace *(far left)* reveals the hexagonal ceramic cores that will give the Columbus mirrors their honeycombed interior. As the furnace revolves *(near left)*, eighteen tons of borosilicate glass will be heated to more than 2,000 degrees Fahrenheit. The spin will force the molten glass out toward the perimeter to produce a deeply curved dish *(below)*.

The cooled 8-meter parabolic blank will vary from 85 centimeters (33.5 inches) thick at the perimeter to 43 centimeters (17 inches) in the center, with its front plate about 3 centimeters (1.25 inches) thick. After a high-pressure water bath flushes out the ceramic mold, the blank will be ready for polishing to a smoothness that allows no bumps or depressions larger than a few millionths of an inch.

FOUR FLEXIBLE EYES

The designers of the Very Large Telescope—four 8-meter (315-inch) mirrors to be housed in four separate structures atop a mountain in Chile—opted for a completely different approach to maintaining the shape of their large reflectors. Rather than spin-casting a rigid honeycombed structure, they have chosen to make their mirrors remarkably flexible. Only 18 centimeters (7 inches) thick—about half the depth of the Columbus reflectors at their thinnest point—the mirrors of the VLT will be kept in shape by a system of constant manipulation and feedback known as active optics. Computer-controlled devices called actuators will push and pull on the back of each mirror to smooth out deformations caused by wind and gravity. When operating in concert, the four mirrors will have the light-gathering power of one 16-meter (630-inch) mirror.

In addition to its active-optics system, the VLT will employ adaptive optics *(opposite)* to counteract the distortions that occur when light waves pass through the atmosphere. About a hundred times more responsive than the active system, the adaptive version is intended to cycle through its entire feedback and correction loop as fast as 300 times per second.

At first, the technique will be used only at infrared wavelengths, which are slightly longer than visible light; the adjustments therefore need not be as precise as for visible light.

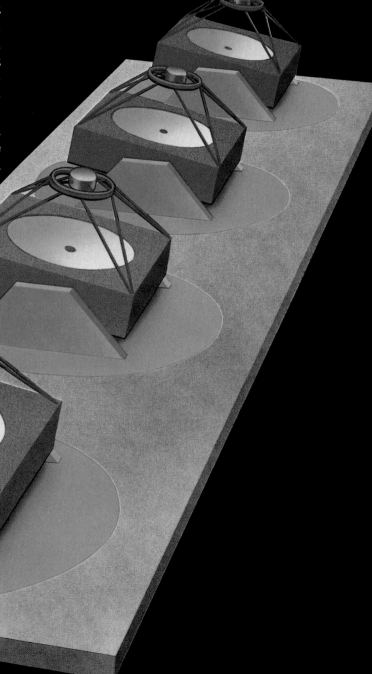

With four 8-meter mirrors arranged on a baseline of about 100 meters, the European Southern Observatory's Very Large Telescope is the largest of the new-technology telescopes planned for the 1990s.

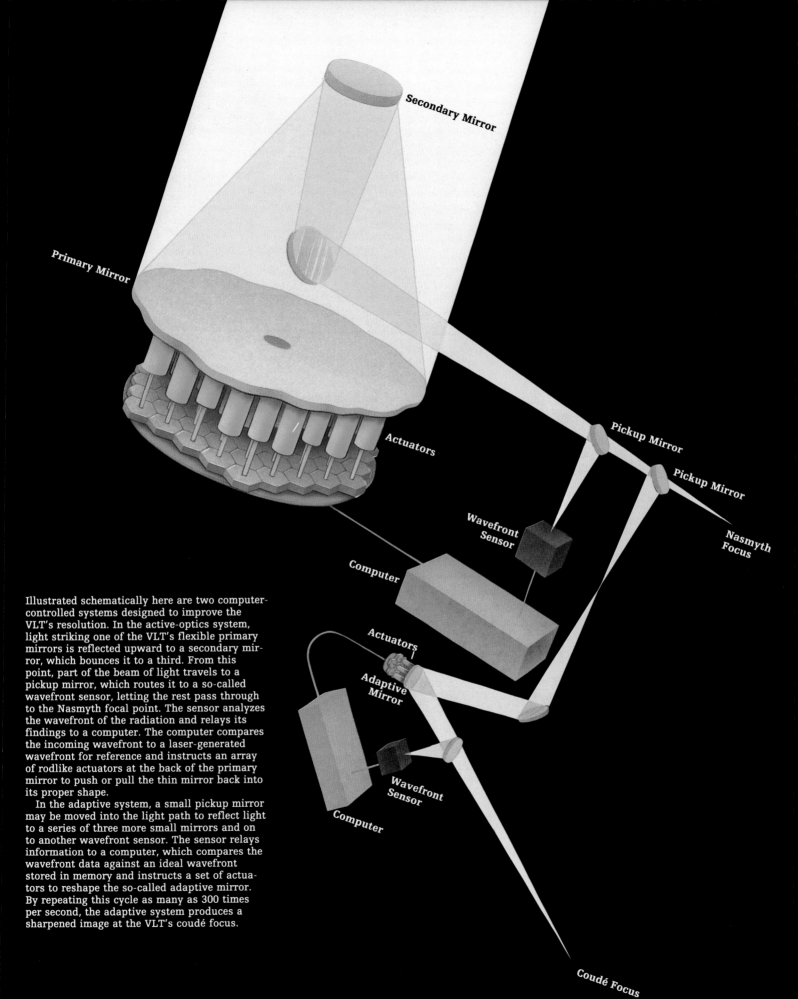

Secondary Mirror

Primary Mirror

Actuators

Pickup Mirror

Pickup Mirror

Wavefront Sensor

Nasmyth Focus

Computer

Actuators

Adaptive Mirror

Wavefront Sensor

Computer

Coudé Focus

Illustrated schematically here are two computer-controlled systems designed to improve the VLT's resolution. In the active-optics system, light striking one of the VLT's flexible primary mirrors is reflected upward to a secondary mirror, which bounces it to a third. From this point, part of the beam of light travels to a pickup mirror, which routes it to a so-called wavefront sensor, letting the rest pass through to the Nasmyth focal point. The sensor analyzes the wavefront of the radiation and relays its findings to a computer. The computer compares the incoming wavefront to a laser-generated wavefront for reference and instructs an array of rodlike actuators at the back of the primary mirror to push or pull the thin mirror back into its proper shape.

In the adaptive system, a small pickup mirror may be moved into the light path to reflect light to a series of three more small mirrors and on to another wavefront sensor. The sensor relays information to a computer, which compares the wavefront data against an ideal wavefront stored in memory and instructs a set of actuators to reshape the so-called adaptive mirror. By repeating this cycle as many as 300 times per second, the adaptive system produces a sharpened image at the VLT's coudé focus.

BUILDING A
MOSAIC MIRROR

Perched at the summit of 13,796-foot-high Mauna Kea in Hawaii, the Keck Telescope is designed to have four times the light-gathering power of the Hale. Its mirror, in contrast to the monolithic reflectors designed for the Columbus and the VLT, is made up of thirty-six segments, cast and shaped individually and then fitted together to form a mosaic 10 meters (394 inches) in diameter. Small segments are much less expensive to manufacture than a single large mirror and offer other practical benefits as well. For example, mirrors need roughly annual resurfacing with reflective aluminum; the Keck's pieces can be removed and treated one at a time, and since the telescope will have spare segments to swap for those being removed, the telescope need never be out of service. Similarly, if a segment is damaged, it is easily replaced.

The Keck needs six each of six differently configured, hexagonal segments to achieve an overall hyperbolic shape for the composite surface. (A hyperbola is a deep, nonspherical curve that, like a parabola, reflects light to a sharp focus.) To meet the challenge of casting and shaping the six different segment types, the Keck's designers invented a process called stress polishing *(opposite)*. Once in operation, the segments will be kept within a millionth of an inch of perfect alignment by an active control system that makes positional adjustments on the segments at the rate of two times a second, based on readings from 168 positional sensors.

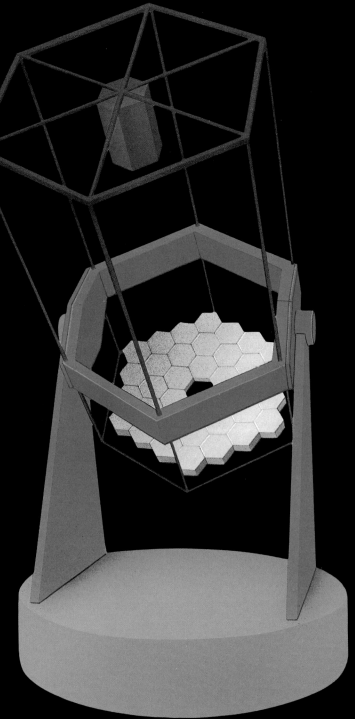

Thirty-six hexagonal segments arranged in three concentric rings make up the 10-meter Keck Telescope. The deep hyperbolic shape of the composite reflector puts prime focus 17.5 meters (57 feet) above the mirror's surface.

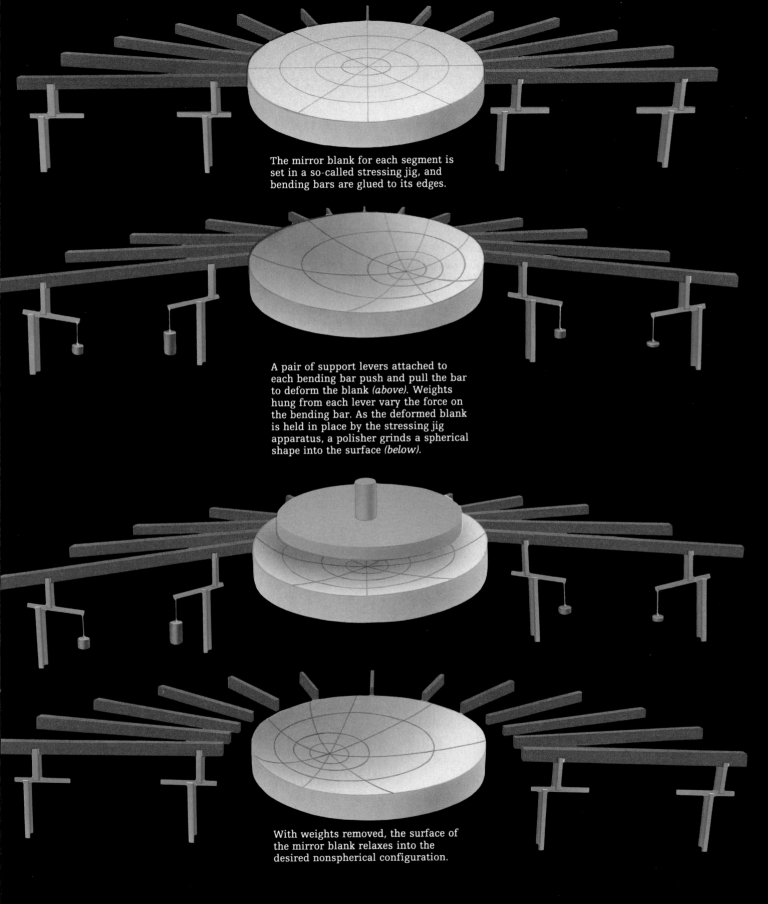

The mirror blank for each segment is set in a so-called stressing jig, and bending bars are glued to its edges.

A pair of support levers attached to each bending bar push and pull the bar to deform the blank *(above)*. Weights hung from each lever vary the force on the bending bar. As the deformed blank is held in place by the stressing jig apparatus, a polisher grinds a spherical shape into the surface *(below)*.

With weights removed, the surface of the mirror blank relaxes into the desired nonspherical configuration.

A Spectral Specialist

The main difference between the Spectroscopic Survey
Telescope and other new telescopes coming off the
drawing board is that the SST is designed to do only
one thing—analyze light that has been broken up into
its constituent wavelengths. Conceived at Pennsylvania State University, the SST is to be housed at the
University of Texas McDonald Observatory, about 150
miles southeast of El Paso. Like the Keck, the SST will
be a mosaic, made up of eighty-five 1-meter (39.4-inch)
segments that will form a primary with a diameter of
8 meters. However, the SST's segments will be identical segments of a sphere—a shape much easier to
grind; the resulting composite surface will form a
spherical curve. The chief drawback of a spherical
reflecting surface is that it does not bring the light that
strikes all parts of its surface to a common focus,
resulting in blurry images. The designers of the SST
have addressed the problem by adding spherical-
aberration correctors *(opposite)*.

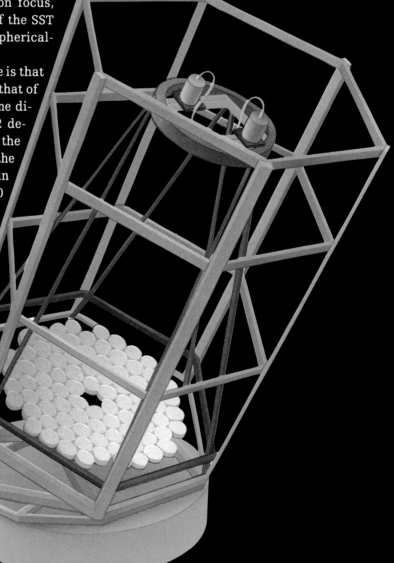

Because it is permanently tilted at 30 degrees
from the vertical, the eighty-five-segment
mirror of the SST will not be subject to changing gravitational pull and will thus not require
an active-optics system.

A singular advantage of a spherical surface is that
it has a field of view many times wider than that of
a parabolic or hyperbolic surface of the same diameter. The SST will capture light from 12 degrees of arc, or about twenty-three times the
angular diameter of the full Moon. One of the
8-meter parabolic mirrors of the Columbus, in
contrast, has a field of view of about only 50
arc seconds. Astronomers will thus be able
to collect spectrographic information from
many stars at once without moving the primary mirror.

To gain a new field of view, the SST
will rotate parallel to the horizon but
not up and down; its primary mirror
will be permanently tilted at an angle
of 30 degrees from the vertical. The
telescope will take advantage of
Earth's rotation to bring 70 percent
of the sky within range. Simplici-
ties like this result in tremendous
economies. With an expected
price tag of well under $10 mil-
lion, the SST will cost just one-
twentieth the amount to be
spent on the VLT.

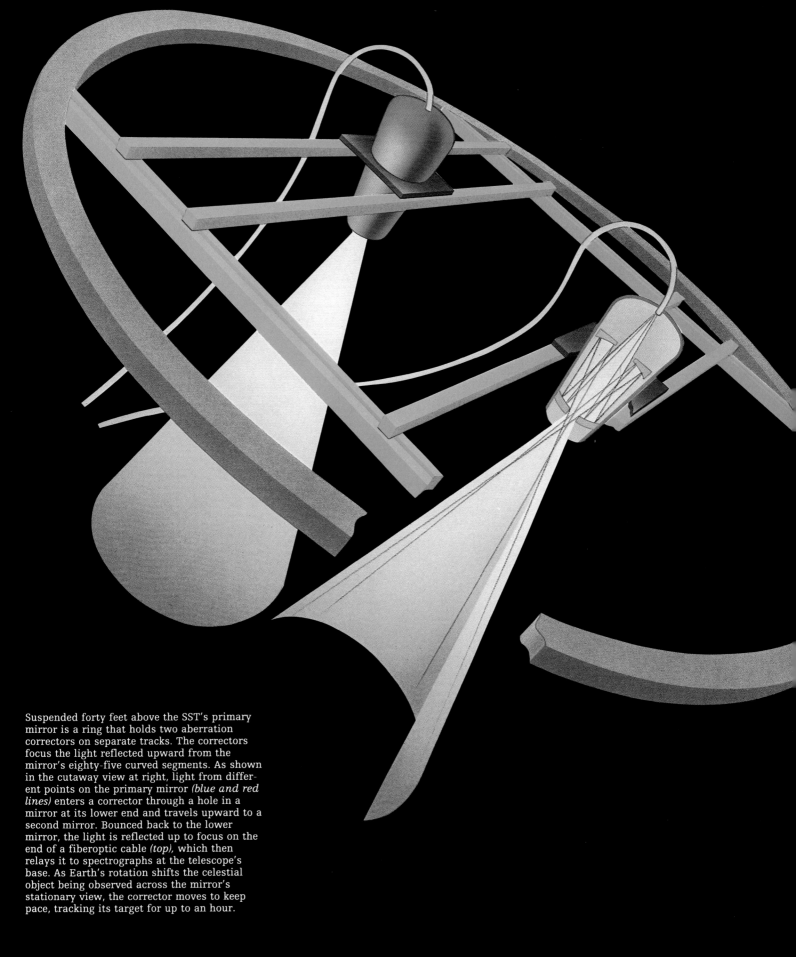

Suspended forty feet above the SST's primary mirror is a ring that holds two aberration correctors on separate tracks. The correctors focus the light reflected upward from the mirror's eighty-five curved segments. As shown in the cutaway view at right, light from different points on the primary mirror *(blue and red lines)* enters a corrector through a hole in a mirror at its lower end and travels upward to a second mirror. Bounced back to the lower mirror, the light is reflected up to focus on the end of a fiberoptic cable *(top),* which then relays it to spectrographs at the telescope's base. As Earth's rotation shifts the celestial object being observed across the mirror's stationary view, the corrector moves to keep pace, tracking its target for up to an hour.

Color added by computer to a black-and-white photo of Halley's comet reveals unsuspected variations in the brilliance of its icy nucleus.

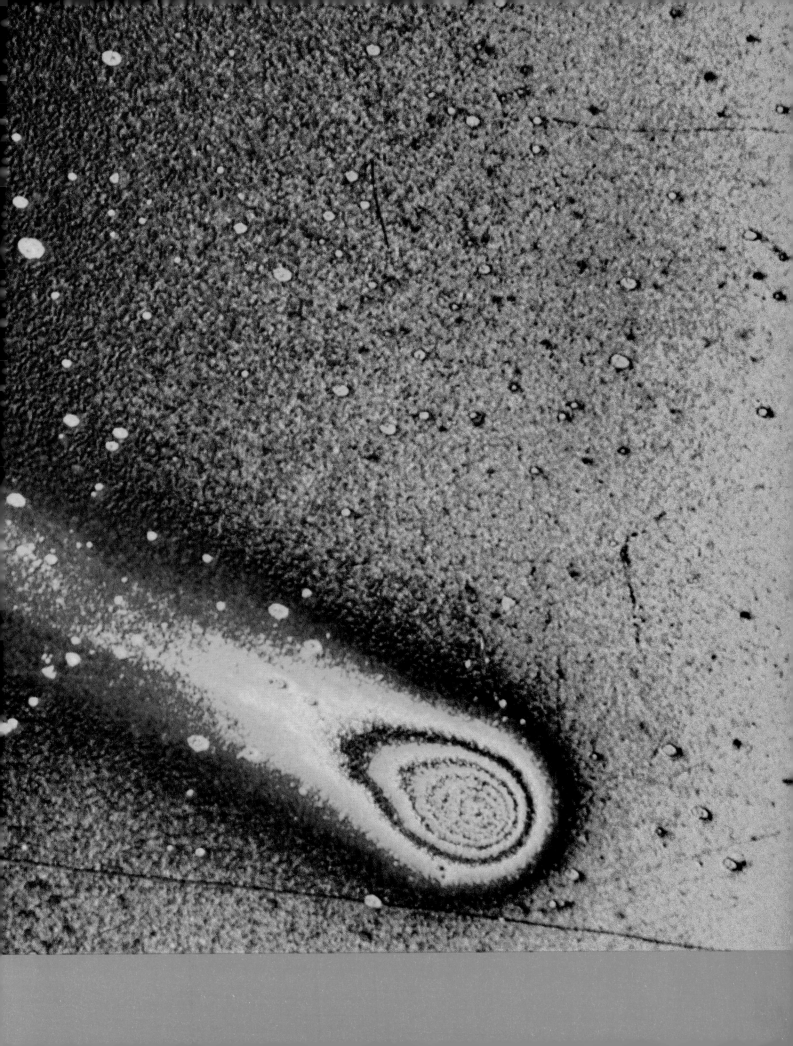

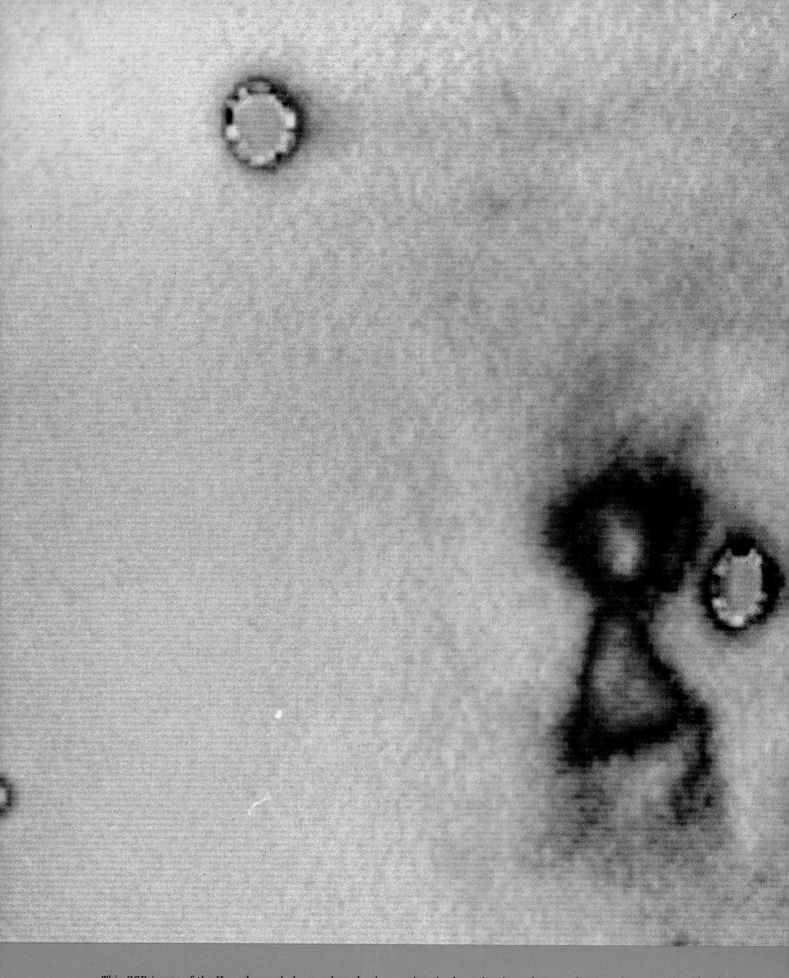

This CCD image of the Hourglass nebula reveals a glowing gas interior heated to incandescence by a newborn star *(nearby ring)*.

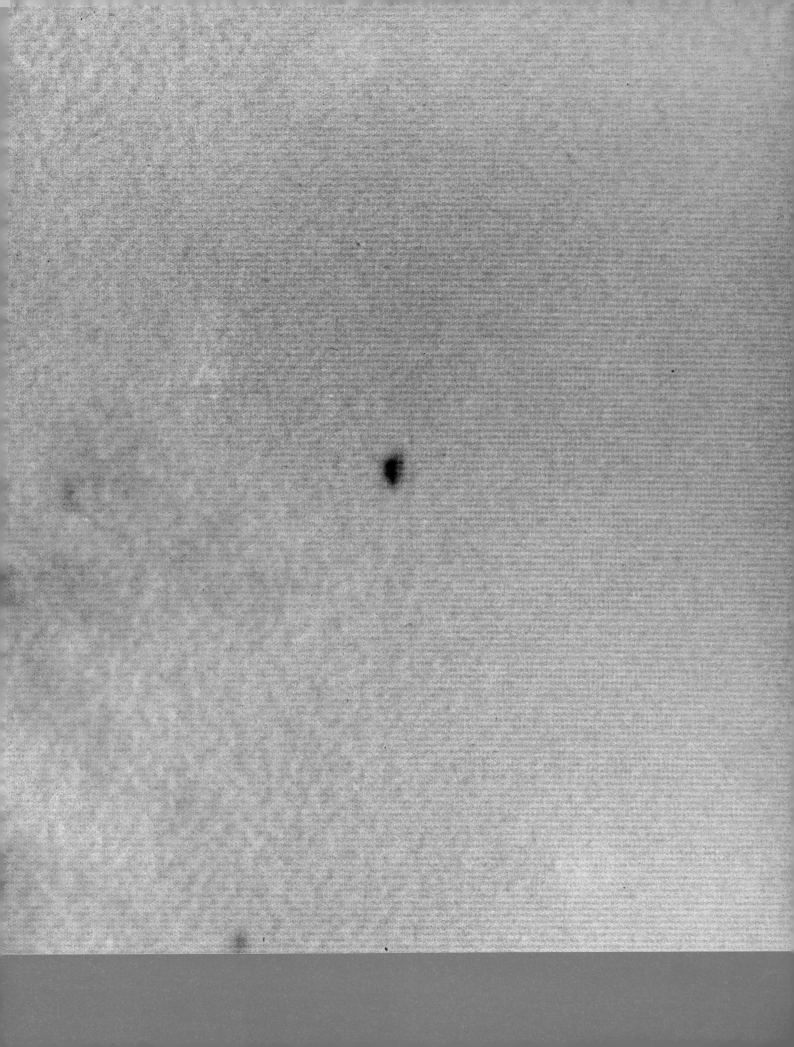

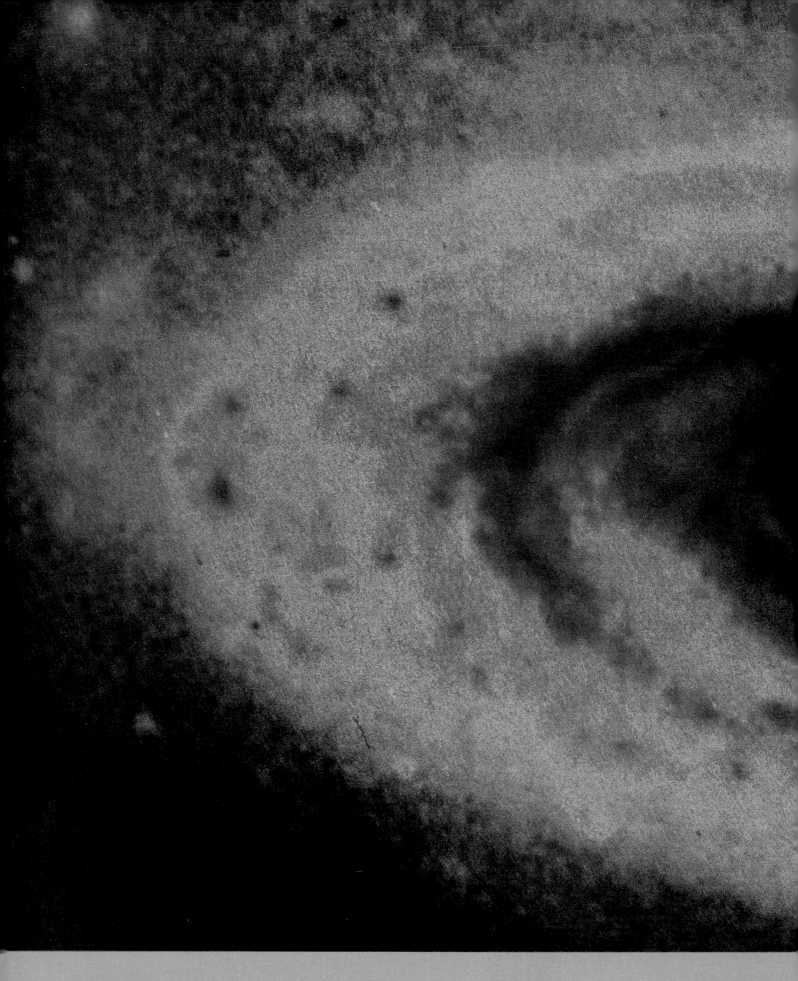

The young, brilliant stars of spiral galaxy NGC 5985 occupy a central yellow ring in this CCD image, colored according to brightness. Older,

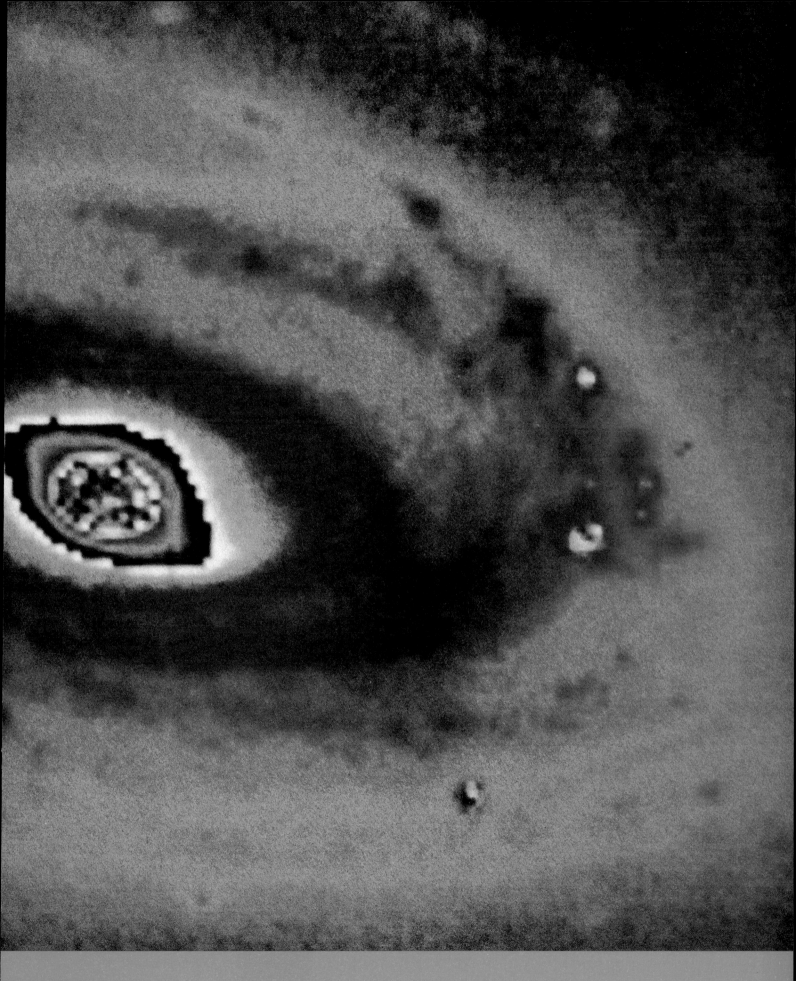

dimmer stars inhabit the green and blue swirls that mark the galaxy's perimeter.

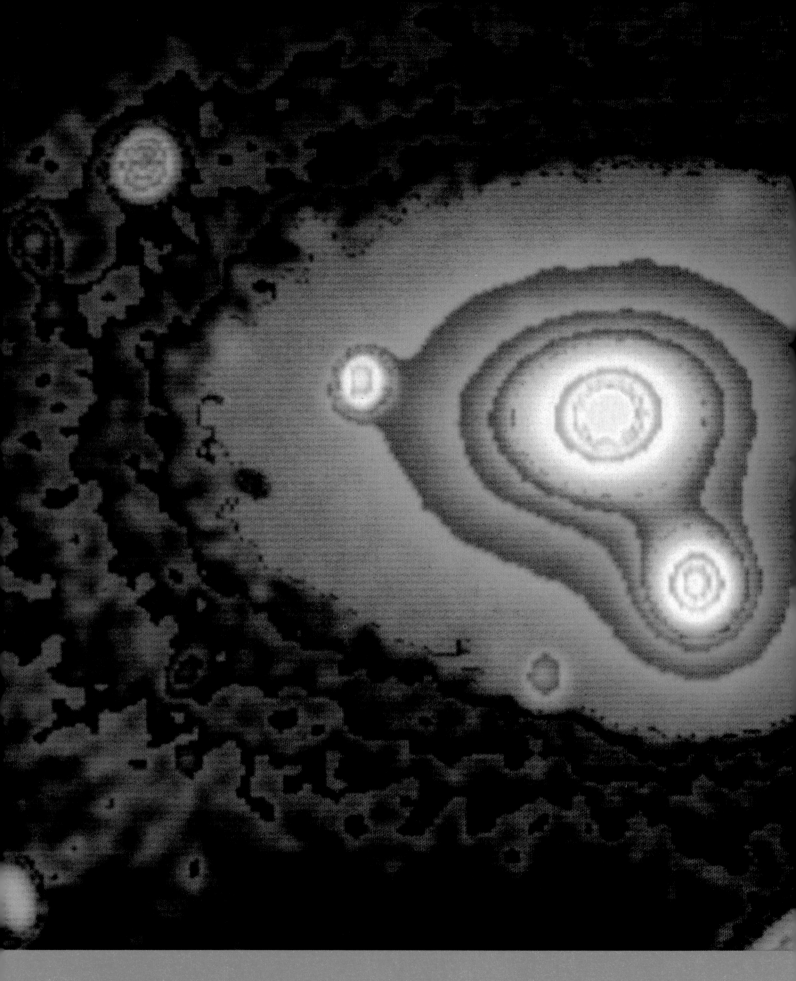

False color assigned to this CCD image shows brightness levels in two prominent galaxies of the cluster Abell 779, which is located 200

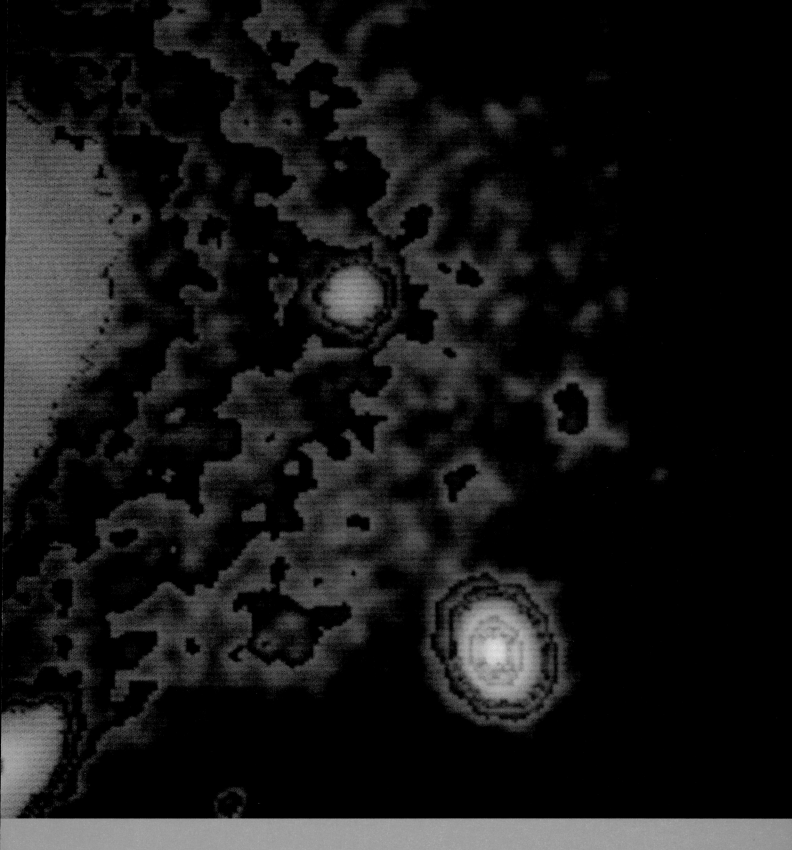

million light-years away in the constellation Cancer.

GLOSSARY

Absorption line: a dark line or band at a particular wavelength in a spectrum, formed when a substance between a radiating source and an observer absorbs the electromagnetic radiation of that wavelength. Different substances produce characteristic absorption lines.

Active optics: a feedback system used with large reflecting telescopes for correcting deformation of the telescope's primary mirror caused by gravity, wind, or changes in temperature. The image reflected by the telescope is relayed to a computer, which compares it to a standard stored in memory. The computer then directs devices known as actuators at the back of the mirror to push or pull the mirror into proper shape or alignment to correct imperfections in the reflected image. These measurements and corrections take place at the rate of one to three times per second.

Adaptive optics: a system used with large reflecting telescopes to correct the distortion that results as light waves pass through so-called convective cells in Earth's atmosphere. Light collected by the primary mirror is directed first to a small, flexible secondary mirror; then, as with active optics, the image is relayed to a computer for comparison with a standard. The computer then directs actuators to adjust the shape of the secondary mirror to compensate for the atmosphere-induced distortion. The adjustments are made to the smaller secondary rather than the large primary mirror, because convective cells shift about so rapidly that corrections must be made at the rate of several hundred times per second.

Altitude: the angular elevation of a celestial object above the observer's horizon.

Altitude over azimuth mount (alt-az mount): a telescope support that allows a telescope to swing vertically in order to find an object's altitude above the horizon, and horizontally in order to find its azimuth, or east-west position. The telescope must be continually adjusted on both axes to keep the object in view. *See* Equatorial mount.

Angular diameter: an object's width on the celestial sphere, measured as an angle in degrees of arc. The Moon's angular diameter is just over half a degree.

Azimuth: an arc along the horizon measured clockwise in degrees between true north and the vertical circle passing through the zenith and the center of the celestial object being observed.

Binary star system: a gravitationally bound pair of stars in orbit around their mutual center of gravity. Binary stars are extremely common, as are systems of three or more stars.

Brightness: the intensity of the light received from an object; a combined result of its actual luminosity, its distance, and light absorption by intervening interstellar dust or gas.

Celestial equator: the projection of Earth's equator onto the celestial sphere.

Celestial horizon: the projection of the observer's horizon onto the celestial sphere.

Celestial sphere: the apparent sphere of sky that surrounds Earth; used by astronomers as a convention for specifying the location of a celestial object.

Charge-coupled device (CCD): an electronic array of detectors, usually positioned at a telescope's focus, for registering electromagnetic radiation.

Chromatic aberration: the tendency of a lens or lens system to bring different wavelengths of light to a focus at different distances from the lens. Because shorter-wavelength blue and violet light tend to focus nearer the lens

than longer-wavelength red light, optical images are characterized by a color halo or other color distortion.

Convective cells: rapidly fluctuating bubbles of turbulence throughout the atmosphere, typically only a few inches in diameter. Electromagnetic radiation passing through the atmosphere to Earth is distorted by convective cells in inverse proportion to its wavelength: the shorter the wavelength, the more the distortion.

Declination: one of two coordinates used to define the position of an object on the celestial sphere. Similar to latitude, an object's declination is measured in degrees north or south of the celestial equator. *See* Right ascension.

Digitization: converting an analog signal—an electrical voltage, for example—into a binary number that can be read by a digital computer and stored in its memory.

Doppler effect: a wave phenomenon in which waves appear to compress to shorter wavelengths as their source approaches the observer or stretch out to longer wavelengths as the source recedes from the observer. *See* Red shift.

Ecliptic: the plane of Earth's orbit around the Sun; also, the apparent path of the Sun across the celestial sphere.

Electromagnetic radiation: waves of electric and magnetic energy that travel through space at 186,000 miles per second.

Electromagnetic spectrum: the continuum of electromagnetic radiation in order of frequency and wavelength, from low-frequency, long-wavelength radio waves to high-frequency, short-wavelength gamma rays.

Electron: a negatively charged particle that normally orbits an atom's positively charged nucleus but may exist as a free particle in extreme low-density conditions.

Emission line: a bright band at a particular wavelength in a spectrum, emitted by hot gases in the source and indicating by its wavelength a chemical constituent of that source.

Emulsion: a suspension of light-sensitive chemical compounds in a medium such as gelatin, mounted on a glass or celluloid base and used in photography to create an image when exposed to a light source.

Equatorial mount: a telescope support that aligns the telescope's vertical, or polar, axis with Earth's axis of rotation so that the polar axis points to the north celestial pole. Once a star or other object is in the field of view, the telescope rotates around the polar axis to keep pace with the star's apparent westward movement as Earth rotates. *See* Altitude over azimuth mount.

Field of view: portion of the sky visible through a telescope, measured in degrees.

Focal length: distance between the objective lens or primary mirror of a telescope and its focus.

Focus: the point at which parallel incident light rays refracted by a lens or reflected by a mirror converge. A photographic plate or other detector placed at the focal plane of a telescope will record the best image of a distant light source afforded by the telescope.

Frequency: the number of oscillations per second of an electromagnetic or other wave. *See* Wavelength.

Infrared: a region of the electromagnetic spectrum with a lower frequency and longer wavelength than visible red light. Most infrared is absorbed by Earth's atmosphere.

Interferometry: the technique of observing a source of electromagnetic radiation by mixing and correlating the simultaneous signals obtained by two or more separated telescopes. Interferometers produce overlapping wave patterns from the radiation; the patterns are studied to determine the angular structure of the emitting source.

Lens: a piece of transparent material, usually glass, used in optical instruments to form an image by focusing light rays.

Luminosity: an object's total radiant energy output, usually measured in ergs per second.

Nebula: a cloud of interstellar dust or gas, in some cases a supernova remnant or a shell ejected by a star.

Objective: a lens or system of lenses in a refracting telescope that forms an image of an object.

Phase: a point in a wave cycle as measured from crest to crest or trough to trough. Two waves are said to be in phase when their crests or troughs align.

Photoelectric effect: the fact that light at sufficiently high frequencies can knock electrons out of a surface.

Photometer: a device for determining an object's brightness by measuring the intensity of its radiation.

Photomultiplier tube: a device used to observe faint light sources by amplifying the flow of electrons produced when the light strikes the photoelectric surface of the detector. For each incoming photon of light, a million or more secondary electrons flow through the tube to enhance an otherwise immeasurably weak current.

Photon: a packet of electromagnetic energy associated with a specific wavelength or frequency. It behaves as a chargeless and massless particle traveling at the speed of light.

Pixel: short for picture element; one of the thousands of dots that make up a digitized image.

Primary mirror: the light-collecting mirror of a reflecting telescope that forms an image of a celestial object.

Quadrant: an early astronomical instrument used to determine the altitude of stars and planets. It consisted of a quarter-circle divided into fractions of degrees and was equipped with a sighting device and a plumb line or spirit level for fixing the vertical or horizontal.

Quasar: a distant, extremely powerful source of energy, much smaller than a galaxy, that can be thousands of times brighter than a normal galaxy.

Radial velocity: the line-of-sight speed of an object toward or away from an observer.

Red shift: a Doppler effect seen when a radiating source recedes from the observer. The received waves lengthen so that any absorption or emission lines move to lower frequencies from their expected frequencies at rest. In visible light, this shift is toward the red end of the spectrum.

Reflector: a mirror or other surface used to collect and focus radiation; also, a telescope that uses such a light-gathering and -focusing surface.

Refraction: the bending of light as it passes from one medium to another medium that has a different index of refraction—from air to glass, for example.

Refractor: a lens used to focus radiation; also, a telescope using such a focusing system.

Resolution: the degree to which spatial details in an image can be distinguished, or resolved. For any given wavelength, a telescope's resolution generally increases with the diameter of its mirror or aperture.

Right ascension: one of two coordinates used to locate objects on the celestial sphere. Similar to longitude on Earth, right ascension is measured eastward in hours, minutes, and seconds from the point where the celestial equator crosses the ascending node of the ecliptic, or the apparent path of the Sun across the celestial sphere. *See* Declination.

Speckle interferometry: a technique for improving the resolution of astronomical images by making a series of very short exposures—called specklegrams—to compensate for the distorting effect of convective cells in Earth's atmosphere. Hundreds of specklegrams are then processed by computer to yield an image that can reveal an object's size, position, and structure.

Spectrograph: an instrument that spreads visible light or other electromagnetic radiation into a spectrum of wavelengths and records the result photographically or electronically.

Spectroscopy: the study of spectra, including the position and intensity of emission and absorption lines, to learn about the chemical makeup of the sources and the physical conditions that create them.

Spectrum: the array of wavelengths obtained by dispersing light as through a prism; often punctuated by absorption or emission lines.

Spherical aberration: a difference in focal length that varies from the center to the edge of a spherical lens or mirror. Since light waves passing through or reflected from the edge focus at a different location than light waves from the center, spherical aberration in a refracting or reflecting telescope causes astronomical images to be blurred.

Supernova: a stellar explosion that expels all or most of the star's mass and is intensely energetic. The explosion usually marks the ending stage in the evolution of a massive star.

Wavefront: a surface composed at any instant of all the points just reached by a wave of electromagnetic radiation (or other vibrational disturbance) in its propagation through a medium.

Wavelength: the distance from crest to crest, or trough to trough, of an electromagnetic or other wave. Wavelengths are related to frequency: the longer the wavelength, the lower the frequency.

Zenith: the point on the celestial sphere directly above a given observer.

BIBLIOGRAPHY

Books

Abetti, Giorgio, *The History of Astronomy*. Transl. by Betty Burr Abetti. New York: Henry Schuman, 1952.

Asimov, Isaac, *Eyes on the Universe*. Boston: Houghton Mifflin, 1975.

Audouze, J., and G. Israel, eds., *Cambridge Atlas of Astronomy*. New York: Cambridge University Press, 1985.

Bailey, Edwin F., "Herschel as a Telescope Maker." In *Telescopes: How to Make Them and Use Them*, ed. by Thornton Page and Lou Williams Page. New York: Macmillan, 1966.

Barlow, Boris V., *The Astronomical Telescope*. London: Wykeham, 1975.

Bates, Ralph S., "Alvan Clark and Sons." In *Telescopes: How to Make Them and Use Them*, ed. by Thornton

Page and Lou Williams Page. New York: Macmillan, 1966.

Bedini, Silvio A.:
"The Instruments of Galileo Galilei." In *Galileo: Man of Science*, ed. by Ernan McMullan. New York: Basic Books, 1967.
"The Makers of Galileo's Scientific Instruments." In *Atti del Simposio. Internazionale di Storia, Metodologie, Logica e Filosofia della Scienza. Galileo nella Storia e nella Filosfia della Scienza*. Vinci (Florence): Gruppo Italiano di Storia della Scienza, 1967.

Berry, Richard, *Build Your Own Telescope*. New York: Charles Scribner's Sons, 1985.

Brandt, John C., and Stephen P. Maran, eds.:
The New Astronomy and Space Science Reader. San Francisco: W. H. Freeman, 1977.
New Horizons in Astronomy. San Francisco: W. H. Freeman, 1979.

Burbidge, G., and A. Hewitt, *Telescopes for the 1980s*. Palo Alto, Calif.: Annual Reviews, 1981.

Chapman, Robert D., *Discovering Astronomy*. San Francisco: W. H. Freeman, 1978.

Cornell, James, and John Carr, eds., *Infinite Vistas: New Tools for Astronomy*. New York: Charles Scribner's Sons, 1985.

Couper, Heather, and Nigel Henbest, *Space Scientist: Telescopes and Observatories*. New York: Franklin Watts, 1987.

Falk, David S., Dieter R. Brill, and David G. Stork, *Seeing the Light: Optics in Nature, Photography, Color, Vision, and Holography*. New York: Harper & Row, 1986.

Ferris, Timothy:
Coming of Age in the Milky Way. New York: Doubleday, 1988.
Galaxies. New York: Stewart, Tabori & Chang, 1982.

Fleischer, Robert, "Story of the Telescope." In *Telescopes: How to Make Them and Use Them*, ed. by Thornton Page and Lou Williams Page. New York: Macmillan, 1966.

Françon, M., *Laser Speckle and Applications in Optics*. Transl. by Henri H. Arsenault. New York: Academic Press, 1979.

Furness, Caroline E., *An Introduction to the Study of Variable Stars*. Boston: Houghton Mifflin, 1915.

Gingerich, Owen, "The Historical Tension between Astronomical Theory and Observation." In *Revealing the Universe: Prediction and Proof in Astronomy*, ed. by James Cornell and Alan P. Lightman. Cambridge, Mass.: MIT Press, 1982.

Hecht, Eugene, *Optics*. Reading, Mass.: Addison-Wesley, 1987.

Henbest, Nigel, *Observing the Universe*. Oxford, England: Basil Blackwell, 1984.

Henbest, Nigel, and Michael Marten, *The New Astronomy*. Cambridge, England: Cambridge University Press, 1983.

Hoskin, Michael A., *William Herschel and the Construction of the Heavens*. New York: W. W. Norton, 1963.

Jastrow, Robert, and Malcolm H. Thompson, *Astronomy Fundamentals & Frontiers*. New York: John Wiley & Sons, 1984.

Kaufman, William J., *Universe*. New York: W. H. Freeman, 1987.

King, Henry C., *The History of the Telescope*. Cambridge, Mass.: Sky, 1955.

Koestler, Arthur, *The Sleepwalkers: A History of Man's Changing Vision of the Universe*. New York: Grosset & Dunlap, 1963.

Kopal, Zdenek, *Telescopes in Space*. London: Faber & Faber, 1968.

Kuiper, Gerard P., and Barbara M. Middlehurst, eds., *Telescopes*. Chicago: University of Chicago Press, 1960.

Laustsen, Svend, Claus Madsen, and Richard M. West, *Exploring the Southern Sky*. New York: Springer-Verlag, 1987.

Learner, Richard, *Astronomy through the Telescope*. New York: Van Nostrand Reinhold, 1981.

Maran, Stephen P., "The Giant Telescopes." In *Science Year 1990*. New York: World Book, in press.

Meyer-Arendt, Jurgen R., *Introduction to Classical and Modern Optics*. London: Prentice-Hall, 1984.

Miczaika, G. R., and William M. Sinton, *Tools of the Astronomer*. Cambridge, Mass.: Harvard University Press, 1961.

Moore, Patrick:
The Guinness Book of Astronomy: Facts & Feats. London: Guinness Superlatives, 1979.
The Picture History of Astronomy. New York: Grosset & Dunlap, 1972.
Watchers of the Stars: The Scientific Revolution. New York: G. P. Putnam's Sons, 1973.

Moore, Patrick, ed., *The International Encyclopedia of Astronomy*. New York: Orion Books, 1987.

Murchie, Guy, *Music of the Spheres: The Material Universe—from Atom to Quasar, Simply Explained*. New York: Dover, 1967.

Murdin, Paul, and David Allen, *Catalogue of the Universe*. New York: Cambridge University Press, 1979.

Palomar Observatory, *Giants of Palomar*. Mount Palomar, Calif.: California Institute of Technology, 1983.

Preston, Richard, *First Light*. New York: Urania, 1987.

Rapport, Samuel, and Helen Wright, eds., *Astronomy*. New York: New York University Press, 1964.

Ronan, Colin A., *The Practical Astronomer*. London: Roxby Press, 1981.

Time-Life Books, Inc., eds., *Computers and the Cosmos*. Alexandria, Va.: Time-Life Books, 1988.

de Vaucouleurs, Gérard, *Astronomical Photography: From the Daguerreotype to the Electron Camera*. Transl. by R. Wright. New York: Macmillan, 1961.

Wallis, Brad D., and Robert W. Provin, *A Manual of Advanced Celestial Photography*. Cambridge, England: Cambridge University Press, 1988.

Wright, Helen, Joan N. Warnow, and Charles Weiner, eds., *The Legacy of George Ellery Hale*. Cambridge, Mass: MIT Press, 1972.

Periodicals
Andersen, Per H.:
"Astronomers Seek High Resolution." *Physics Today*, June 1987.
"Will Future Astronomers Observe with Liquid Mirrors?" *Physics Today*, June 1987.
"AstroNews: Biggest Spin-Cast Mirror Produced in Arizona." *Astronomy*, October 1988.
"AstroNews: New Technologies for Telescopes." *Astronomy*, February 1989.

Bagnuolo, William G., Jr., and James R. Sowell, "Binary Star Speckle Photometry, I: The Colors and Spectral Types of the Capella Stars." *The Astronomical Journal,* September 1988.

Beatty, J. Kelly:
"HST: Astronomy's Greatest Gambit." *Sky & Telescope,* May 1985.
"Will Space Telescope Be Ready?" *Sky & Telescope,* February 1987.

Bennett, J. A., "The Discovery of Uranus." *Sky & Telescope,* March 1981.

Berry, Richard:
"The Telescope That Defies Gravity." *Astronomy,* July 1988.
"Thinking Telescopes." *Astronomy,* August 1988.

Bracewell, Ronald N., "The Fourier Transform." *Scientific American,* June 1989.

Crawford, David L., "The Ever-Vigilant GNAT." *Sky & Telescope,* February 1989.

di Cicco, Dennis, "Astrophotography Then and Now." *Sky & Telescope,* November 1988.

Dickman, Steven, "Southern Sky Surveyed." *Nature,* April 27, 1989.

Eberhart, J., "More Than Just a Spot: Facing an Asteroid at Last." *Science News,* November 28, 1987.

"Europe to Build World's Largest Optical Telescope." *Astronomy Now,* February 1988.

"Faint Limit of the Universe." *Sky & Telescope,* October 1988.

Fienberg, Richard Tresch, "The New, Improved Space Telescope." *Sky & Telescope,* February 1989.

Fischer, Daniel, "A Telescope for Tomorrow." *Sky & Telescope,* September 1989.

Fisher, Arthur, "Spinning Scopes." *Popular Science,* October 1987.

"Fourier Transformation." *Scientific American,* July 1987.

Gustafson, John R., and Wallace Sargent, "The Keck Observatory: 36 Mirrors Are Better Than One." *Mercury,* March-April 1988.

Helfand, David, "Mystery Spots, X Rays, γ Rays: Is the Dust Settling from SN1987a?" *Physics Today,* January 1988.

Jacchia, Luigi, "Forefathers of the MMT." *Sky & Telescope,* February 1978.

Janesick, James, and Morley Blouke, "Sky on a Chip: The Fabulous CCD." *Sky & Telescope,* September 1987.

Jones, Brian, "William Herschel: Pioneer of the Stars." *Astronomy,* November 1988.

Kanipe, Jeff, "Quest for the Most Distant Objects in the Universe." *Astronomy,* June 1988.

Kristian, Jerome, and Morley Blouke, "Charge-Coupled Devices in Astronomy." *Scientific American,* October 1982.

Labeyrie, Antoine, "Stellar Interferometry Methods." *Annual Review of Astronomy and Astrophysics,* 1978.

McAlister, Harold A.:
"The Apparent Orbit of Capella." *Astronomical Journal,* May 1981.
"Binary-Star Speckle Interferometry." *Sky & Telescope,* May 5, 1977.
"High Angular Resolution Measurements of Stellar Properties." *Annual Review of Astronomy and Astro-physics,* 1985.
"Seeing Stars with Speckle Interferometry." *American Scientist,* March-April 1988.

Maddison, Ron:
"Active and Adaptive Optics." *Astronomy Now,* December 1988.
"The Earliest Telescopes (Part 2): Newton and the First Reflectors." *Astronomy Now,* February 1988.

Malin, David F., "The Developing Art of Star Photography." *New Scientist,* December 17, 1988.

Malin, David F., and William J. Zealey, "Astrophotography with Unsharp Masking." *Sky & Telescope,* April 1979.

Mammana, Dennis L., "The Incredible Spinning Oven." *Sky & Telescope,* July 1985.

Maran, Stephen P., "A New Generation of Giant Eyes Gets Ready to Probe the Universe." *Smithsonian,* June 1987.

Merritt, J. I., "Astrophysical Instrument Maker." *Princeton Alumni Weekly,* September 26, 1984.

Miller, Joseph S., "The Structure of Emission Nebulas." *Scientific American,* October 1974.

Mitton, Simon, "Spin-Casting Magic Spells Big Mirrors." *Astronomy Now,* April 1989.

Murray, Mary, "Spinning Mirrors of Mercury." *Science News,* August 16, 1986.

Müürsepp, P., and E. G. Forbes, "Some Recollections by Contemporaries of Bernhard Schmidt." *Journal of the British Astronomical Association,* December 1969.

"NTT Mirror Completed." *Sky & Telescope,* October 1988.

"NTT Update: Mirror Arrives in Chile." *Astronomy Now,* November 1988.

Nisenson, Peter, and Costas Papaliolios, "Effects of Photon Noise on Speckle Image Reconstruction with the Knox-Thompson Algorithm." *Optics Communications,* August 15, 1983.

Nisenson, Peter, et al., "Data Recording and Processing for Speckle Image Reconstruction." *Applications of Speckle Phenomena,* 1980.

O'Dell, C. Robert, "Building the Hubble Space Telescope." *Sky & Telescope,* July 1989.

Papaliolios, Costas, Peter Nisenson, and Steve Ebstein, "Speckle Imaging with the PAPA Detector." *Applied Optics,* January 15, 1985.

Readhead, Anthony C. S., "Radio Astronomy by Very-Long-Baseline Interferometry." *Scientific American,* June 1982.

Robinson, Leif J., "Monster Telescopes for the 1990's." *Sky & Telescope,* May 1987.

"Schott to Spin-Cast Giant Mirrors." *Sky & Telescope,* November 1988.

Schultz, Ron, "Stars." *Omni,* October 1988.

Sinnott, Roger W., "Gleanings for ATM's." *Sky & Telescope,* January 1982.

"Speckled Vesta." *Sky & Telescope,* June 1987.

Stobie, R. S., R. J. Dodd, and H. T. MacGillivray, "Image Processing with COSMOS." *Sky & Telescope,* December 1981.

Sulentic, Jack W., and Jean J. Lorre, "The Magic of Image Processing." *Sky & Telescope,* May 1984.

Thomsen, Dietrick E.:
"Big Telescopes on a Roll." *Science News,* September 12, 1987.

"Taking the Measure of the Stars." *Science News,* January 3, 1987.

Tobin, William, "Perfecting the Modern Reflector." *Sky & Telescope,* October 1987.

Trefill, James, "From Astronomy to Astrophysics." *Wilson Quarterly,* Summer 1987.

Waldrop, M. Mitchell:
"Keck Telescope Mirror Is in Production." *Science,* February 24, 1989.
"NASA's $60,000 Epoxy Drops." *Science,* March 10, 1989.
"The New Eye of Texas Is Soon to Be upon Us." *Science,* February 19, 1988.
"Seeing All There Is to See in the Universe." *Science,* July 22, 1988.
"Telescope Gets Largest Private Gift Ever." *Science,* January 18, 1985.
"Will the Hubble Space Telescope Compute?" *Research News,* March 17, 1989.

West, Richard M., "Europe's Astronomy Machine." *Sky & Telescope,* May 1988.

Other Sources

Brandt, John C., "Progress Report on the High Resolution Spectrograph for the Space Telescope." In *Ultraviolet and Vacuum Ultraviolet Systems* (Proceedings of SPIE—The International Society for Optical Engineering, Vol. 279). Bellingham, Wash.: SPIE, 1981.

"The CCD: A Versatile Electronic Device Comes of Age." Murray Hill, N.J.: Bell Telephone Laboratories, no date.

NASA Scientific and Technical Information Branch, "The Space Telescope Observatory." Special session of Commission 44, IAU 18th General Assembly, Patras, Greece, August 1982.

"Very Large Telescopes and Their Instrumentation." ESO conference, Garching, March 21-24, 1988.

INDEX

ACKNOWLEDGMENTS

The editors of *The Visible Universe* wish to thank these in-
dividuals for their valuable contributions: Dana Berry,
Space Telescope Science Institute, Baltimore; William
Blair, The Johns Hopkins University, Baltimore; Jonathan
Bland, Rice University, Houston; Daniel K. Brocious, Fred
Lawrence Whipple Observatory, Amado, Ariz.; Alain Bucher,
Observatoire du Pic-du-Midi, France; Georgio Buonvio,
Osservatorio Astronomico, Rome; Nathaniel P. Carleton,
Smithsonian Observatory, Cambridge, Mass.; Matt
Cheselka, University of Arizona, Tucson; Geoff Chester,
National Air and Space Museum, Smithsonian Institution,
Washington, D.C.; Ed Collins, Perkin-Elmer Corporation,
Norwalk, Conn.; Walter Davison, University of Arizona, Tuc-
son; Steven J. Dick, U.S. Naval Observatory, Washington,
D.C.; Richard Dreiser, Yerkes Observatory, Williams Bay,
Wis.; Hans Elsasser, Max Planck Institut für Astronomie,
Heidelberg, Germany; Harland Epps, University of Cali-
fornia, Santa Cruz; Sandra Faber, University of California,
Santa Cruz; Terry Facey, Perkin-Elmer Corporation, Nor-
walk, Conn.; Robert Futually, Observatoire Midi-Pyrénées,
France; John Gustafson, California Association for Research
in Astronomy, Oakland; Robert S. Harrington, U.S. Naval
Observatory, Washington, D.C.; E. Keith Hege, University of
Arizona, Tucson; John Hill, University of Arizona, Tucson;

Peter Hingley, Royal Astronomical Society, London; James Janesick, Jet Propulsion Laboratory, Pasadena, Calif.; Margarita Karovska, Harvard-Smithsonian Center for Astrophysics, Cambridge, Mass.; Arnold R. Klemola, University of California, Santa Cruz; John Lorre, Jet Propulsion Laboratory, Pasadena, Calif.; Bruno Marano, Osservatorio Astronomico dell'Università di Bologna; Terry Mast, Keck Observatory Science Office, Berkeley, Calif.; Fritz Merkle, European Southern Observatory, Garching, Germany; Richard Muller, Observatoire du Pic-du-Midi, France; Domenico Nanni, Osservatorio Astronomico, Monte Porzio, Catone, Italy; Peter Nisenson, Harvard-Smithsonian Center for Astrophysics, Cambridge, Mass.; Jacques-Clair Noëns, Observatoire du Pic-du-Midi, France; Costas Papaliolios, Harvard-Smithsonian Center for Astrophysics, Cambridge, Mass.; Axel Quetsch, Max Planck Institute für Astronomie, Heidelberg, Germany; Larry Ramsey, University Park, Pa.; Janet Sandland, Royal Observatory, Edinburgh; Antonello Satta, Osservatorio Astronomico, Padua, Italy; Nigel Sharp, National Optical Astronomy Observatories, Tucson, Ariz.; Lyman Spitzer, Princeton, N.J.; Hans Vehrenburg, Dusseldorf, Germany; P. Véron, Observatoire de Haute-Provence, France; Ray Villard, Space Telescope Science Institute, Baltimore; Marie-José Vin, Observatoire de Haute-Provence, France; Gerd Weigelt, Max Planck Institut für Radioastronomie, Bonn, Germany; Richard M. West, European Southern Observatory, Garching, Germany; Ray Wilson, European Southern Observatory, Garching, Germany; Marina Zuccoli, Osservatorio Astronomico dell'Università di Bologna, Italy.

PICTURE CREDITS

The sources for the illustrations in this book are listed below. Credits from left to right are separated by semicolons; credits from top to bottom are separated by dashes.

Cover: Art by Matt McMullen. 2, 3: NASA, LBJ Space Center, Houston (51B-116-005). 4: © Roger Ressmeyer, Starlight. 5: CNRS-Observatoire de Haute-Provence. 6, 7: Osservatorio Astronomico, Dipartimento di Astronomia dell'Università, Padua. 8, 9: Jean Lorre, SPL, London. 10, 11: Royal Observatory, Edinburgh. 16, 17: © Roger Ressmeyer, Starlight. 18: Initial cap, detail from pages 16, 17. 20, 21: Art by Fred Holz. 23: Art by Stephen Wagner. 24, 25: Archiv für Kunst und Geschichte, Berlin; Scala/Art Resource, New York (no. K-52146); Archiv für Kunst und Geschichte, Berlin (2); Ann Ronan Picture Library, Taunton, Somerset; The Science Museum, London. Background art by Stephen Wagner. 26, 27: Bausch & Lomb; reproduced by kind permission of the President and Council of The Royal Society, London; The Science Museum, London; courtesy U.S. Naval Observatory Library, photographed by Larry Sherer; The Science Museum, London (2). Background art by Stephen Wagner. 28, 29: Reproduced by kind permission of the President and Council of The Royal Society, London; Stuart Eadon-Allen, Birmingham, Ala.; Yerkes Observatory photograph, University of Chicago; Royal Astronomical Society Library, London; courtesy U.S. Naval Observatory Library, photographed by Larry Sherer—Novosti photograph by V. Maask; Yerkes Observatory photograph, University of Chicago; Royal Astronomical Society Library, London. Background art by Stephen Wagner. 30, 31: Harvard College Observatory (2); Yerkes Observatory, University of Chicago (2); photograph courtesy Mount Wilson Institute, Pasadena, Calif. (2); Hale Observatories, Pasadena, Calif. (2). Background art by Stephen Wagner. 33-39: Art by Stephen Wagner. 40, 41: Art by Fred Holz. 42, 43: Yerkes Observatory, University of Chicago. 44: Royal Astronomical Society, London. 47: National Maritime Museum, London. 50, 51: The Observatories of the Carnegie Institution of Washington. 52-63: Photographs courtesy California Institute of Technology, Palomar Observatory, Pasadena, Calif. Art by Fred Holz. 64, 65: Photo by H. Pedersen, courtesy European Southern Observatory, from *Exploring the Sky,* by Svend Lausten, Claus Madsen, and Richard M. West, Springer Verlag. 66: Initial cap, detail from pages 64, 65. 68: Harvard College Observatory—courtesy D. Scott Birney, Department of Astronomy, Wellesley College, photographed by Larry Sherer. 71: Bibliothèque Nationale, Paris. 72, 73: Hamburger Sternwarte, courtesy A. Weigert. 75: Royal Observatory, Edinburgh, and Anglo-Australian Telescope Board—© 1981 Anglo-American Telescope Board. 76-79: Photographs courtesy Harold A. McAlister, Center for High Angular Resolution Astronomy (CHARA), Georgia State University. Art by Alfred Kamajian. 80, 81: Photographs courtesy Keith Hege and Matt Cheselka, Steward Observatory, University of Arizona. Art by Alfred Kamajian. 82: Keith Hege and Matt Cheselka, Steward Observatory, University of Arizona (2)—K. Hofmann, W. Mauder, D. Schertl, G. Weigelt, Max Planck Institut für Radioastronomie, Bonn (2). 83: K. Hofmann, W. Mauder, D. Schertl, G. Weigelt, Max Planck Institut für Radioastronomie (2)—Peter Nisenson, Costas Papaliolios, Clive Standley, Margarita Karovska, Harvard-Smithsonian Center for Astrophysics, and Steven Heathcote, Cerro Tololo Inter-American Observatory, La Serena, Chile (2). 86: J. Janesick, California Institute of Technology, Pasadena, Calif. 89: Courtesy James Gunn. 91-99: Art by Al Kettler. 94: National Optical Astronomy Observatories, Kitt Peak National Observatory, Tucson. 95: William Blair, Center for Astronomical Sciences, The Johns Hopkins University—California Institute of Technology, Palomar Observatory—Jack W. Sulentic and William C. Keel, University of Alabama (2). 96: Jonathan Bland and R. Brent Tully, Institute for Astronomy, Honolulu—Smithsonian Astrophysical Observatory, Cambridge, Mass. 97: Photographs courtesy Harvard-Smithsonian Center for Astrophysics. 98, 99: Photographs courtesy Jet Propulsion Lab, Pasadena, Calif. 100, 101: European Southern Observatory (2). 102: Initial cap, detail from pages 100, 101. 105: MMT Observatory, Amado, Ariz. 107: NASA. 108: NASA/J. Westphal—NASA/ESA. 109: NASA. 110: Courtesy National Research Council, Canada, photographed by René Racine; NASA/ESA. 113: Dipartimento di Astronomia dell'Università, Bologna. 114: Lori Stiles, University of Arizona. 116: European Southern Observatory, Garching. 117: Courtesy Ray Wilson and European Southern Observatory. 120-127: Art by Matt McMullen. 128, 129: Geoff Chester. 130, 131: Royal Greenwich Observatory, SPL, London. 132, 133: Royal Greenwich Observatory, SPL, London. 134, 135: O.M.P., Toulouse, France.

Time-Life Books is a division of Time Life Inc.,
a wholly owned subsidiary of
THE TIME INC. BOOK COMPANY

TIME-LIFE BOOKS

PRESIDENT: Mary N. Davis
Managing Editor: Thomas H. Flaherty
Director of Editorial Resources:
Elise D. Ritter-Clough
Director of Photography and Research:
John Conrad Weiser
Editorial Board: Dale M. Brown, Roberta Conlan,
Laura Foreman, Lee Hassig, Jim Hicks, Blaine
Marshall, Rita Thievon Mullin, Henry Woodhead

PUBLISHER: Robert H. Smith

Associate Publisher: Trevor Lunn
Editorial Director: Donia Steele
Marketing Director: Regina Hall
Production Manager: Marlene Zack
Supervisor of Quality Control: James King

Editorial Operations
Production: Celia Beattie
Library: Louise D. Forstall

Computer Composition: Deborah G. Tait
(Manager), Monika D. Thayer, Janet Barnes
Syring, Lillian Daniels

Correspondents: Elisabeth Kraemer-Singh (Bonn),
Christina Lieberman (New York), Maria Vincenza
Aloisi (Paris), Ann Natanson (Rome). Valuable
assistance was also provided by Trini Bandres
(Madrid); Dick Berry (Tokyo); John Dunn
(Melbourne); Barbara Hicks, Christine Hinze
(London); Sasha Isachenko (Moscow); Robert
Kroon (Geneva); Angie Lemmer (Bonn); Patricia
Strathern (Paris); Wibo Van de Linde (Amster-
dam); and Ann Wise (Rome).

VOYAGE THROUGH THE UNIVERSE

SERIES DIRECTOR: Roberta Conlan
Series Administrators: Judith W. Shanks,
Susan Stuck

Editorial Staff for *The Visible Universe*
Designers: Cynthia T. Richardson (principal),
Dale Pollekoff
Associate Editor: Tina S. McDowell (pictures)
Text Editors: Lee Hassig (principal),
Allan Fallow
Researchers: Katya Sharpe Cooke,
Edward O. Marshall, Barbara Sause
Writers: Robert M. S. Somerville, Stephanie Lewis
Assistant Designer: Brook Mowrey
Copy Coordinator: Darcie Conner Johnston
Picture Coordinator: Ruth Moss
Editorial Assistant: Katie Mahaffey

Special Contributors: Sarah Brash, K. C. Cole,
James Dawson, Stephen Hart, Steve Maran, Gina
Maranto, Peter Pocock, Chuck Smith (text);
Julianne L. Baum, Adam Dennis, Edward Dixon,
Eugenia Scharf, Jacqueline L. Shaffer, Elizabeth
Thompson (research); Barbara L. Klein (index).

CONSULTANTS

ROBERT J. BRUCATO is assistant director of the
Palomar Observatory of the California Institute of
Technology in Pasadena, California.

JAMES E. GUNN has invented, built, and modified
many observing instruments while also working as
an observing astronomer. He is on the faculty of
Princeton University.

ROBERT LATHAM teaches mathematics, science,
and technology at the Jefferson High School for
Science and Technology, Alexandria, Virginia,
where he is director of the Optics and Modern Physics Laboratory.

HAROLD A. McALISTER is professor of physics and
astronomy at Georgia State University, Atlanta, and
is director of its Center for High Angular Resolution
Astronomy. He has conducted extensive research in
binary star speckle interferometry.

STEPHEN P. MARAN, a senior staff scientist at
NASA Goddard Space Flight Center, writes widely in
astronomy and is press officer for the American
Astronomical Society.

HELEN MORTENSEN is an image processing ana-
lyst and programmer for the Voyager Project at the
Jet Propulsion Laboratory of the California Institute of Technology.

ROBERT W. SMITH is a historian of astronomy who
holds joint appointments at the National Air and
Space Museum, Smithsonian Institution, and in the
History of Science Department of the Johns Hopkins
University.

JACK SULENTIC is a professor of astronomy at the
University of Alabama, Tuscaloosa. He specializes
in image processing and studies of the distance
scale of the universe.

DANIEL WEEDMAN, who teaches at the Pennsylvania State University at University Park, conceived
the Spectroscopic Survey Telescope, which is to be
built in Fort Davis, Texas, in the 1990s. The SST is
a joint project of Penn State and the University of
Texas.

Library of Congress Cataloging in
Publication Data
The Visible universe/by the editors of
Time-Life Books.—Rev. Ed.
   p. cm.   —(Voyage through the universe).
   Includes bibliographical references and
   index.
   ISBN 0-8094-9050-1
   ISBN 0-8094-9051-X (lib. bdg.)
   1. Astronomy. I. Time-Life Books.
   II. Series.
QB44.2.V57 1991
522—dc20   91—36740
                              CIP

For information on and a full description of
any of the Time-Life Books series, please call
1-800-621-7026 or write:
Reader Information
Time-Life Customer Service
P.O. Box C-32068
Richmond, Virginia 23261-2068

Revisions Staff

EDITOR: Roberta Conlan

Associate Editor/Research: Quentin G. Story
Art Director: Robert K. Herndon
Assistant Art Director: Kathleen Mallow
Copy Coordinator: Juli Duncan
Picture Coordinator: David Beard

Consultant: Stephen P. Maran. *See consultants.*

Special thanks to Ray Villard and the Space
Telescope Science Institute, Baltimore.

Earth: diameter 7,926 miles

Neptune: diameter 30,775 miles

Uranus: diameter 31,763 miles

Red supergiant: diameter 400 million miles

Solar System: diameter 7.5 billion miles

Globular cluster: diameter 2×10^{14} miles

Milky Way: diameter 100,000 light-years

Local Group of galaxies:
6 million light-years across

Largest double radio source:
length 17 million light-years